STUMBLING BLOCKS
AGAINST UNIFICATION
On Some Persistent Misconceptions in Physics

Other Related Titles from World Scientific

Mathematical Foundations of Quantum Field Theory
by Albert Schwarz
ISBN: 978-981-3278-63-9

Probing Particle Physics with Neutrino Telescopes
edited by Carlos Pérez de los Heros
ISBN: 978-981-3275-01-0

Quantum Field Theory II
by Mikhail Shifman
ISBN: 978-981-3234-18-5

Field Theory: A Path Integral Approach
Third Edition
by Ashok Das
ISBN: 978-981-120-254-4
ISBN: 978-981-120-266-7 (pbk)

STUMBLING BLOCKS AGAINST UNIFICATION

On Some Persistent Misconceptions in Physics

Matej Pavšič

Jožef Stefan Institute, Slovenia

World Scientific

NEW JERSEY · LONDON · SINGAPORE · BEIJING · SHANGHAI · HONG KONG · TAIPEI · CHENNAI · TOKYO

Published by

World Scientific Publishing Co. Pte. Ltd.
5 Toh Tuck Link, Singapore 596224
USA office: 27 Warren Street, Suite 401-402, Hackensack, NJ 07601
UK office: 57 Shelton Street, Covent Garden, London WC2H 9HE

British Library Cataloguing-in-Publication Data
A catalogue record for this book is available from the British Library.

STUMBLING BLOCKS AGAINST UNIFICATION
On Some Persistent Misconceptions in Physics

ISBN 978-981-121-700-5 (hardcover)
ISBN 978-981-121-701-2 (ebook for institutions)
ISBN 978-981-121-702-9 (ebook for individuals)

For any available supplementary material, please visit
https://www.worldscientific.com/worldscibooks/10.1142/11738#t=suppl

Desk Editor: Ng Kah Fee

Typeset by Stallion Press
Email: enquiries@stallionpress.com

Printed in Singapore

To my wife Mojca
and
daughters Tjaša and Katja

Preface

In recent years, physicists have been facing increasing discomfort due to the frustrating situation in which fundamental physics currently finds itself. As stated by John Baez in a recent book review [1], this discomfort is reflected in the recent book *Foundations of Mathematics and Physics One Century After Hilbert: New Perspectives*, edited by Joseph Kouneiher (Springer, 2018). "The problem is that the progress, extremely rapid during most of the twentieth century, has greatly slowed since the 1970s", says Baez. In fact, as noted by Lee Smolin in his book *The Trouble with Physics* [1], during the last forty years, there have been no frontier-breaking advances in fundamental theoretical physics. In this book, I will explain my view on why this lack of progress exists. At the beginning of the twenty-first century, I published a book titled *The Landscape of Theoretical Physics: A Global View; From Point Particle to the Brane World and Beyond, in Search of a Unifying Principle* (Kluwer Academic, 2001). At that time, I saw the problem that many directions of research are pursued more or less independently from each other. Therefore, from my perspective, I presented a global view of the connections and interrelations among those various fields under investigation.

Since then, I have realized that certain points considered in my book, and some other points that I investigated afterward, belonged to the topics about which the majority of physicists already had their fixed opinions. I realized that such fixed opinions were rooted in the existing paradigms, which currently act as stumbling blocks of further genuine progress in fundamental (theoretical) physics. It has become increasingly clear to me that an extensive exposition of those problematic topics is necessary to help us step out of such conceptual boxes.

---

[1] Notices of the American Mathematical Society, Vol. 66, Number 10, p. 1690 (2019).

In the history of physics and science in general, we encounter numerous misconceptions about how Nature works. The situation nowadays certainly is not very different. In the book, I will discuss why the general opinion about the devastating role of negative energies in classical and quantum physics, especially in quantum field theories, should be revised for the benefit of further advances in quantum gravity. To determine under which conditions a physical system is stable or unstable is not so straightforward as it is generally uncritically believed. Once this concept is understood, the doors are opened to welcoming ultrahyperbolic spaces, whose signature has positive and negative signs. Such spaces are associated with Clifford algebras, which have numerous promising applications, as will be discussed throughout this book. Among others, the generators of Clifford algebras in infinite-dimensional spaces play the role of quantum field operators and create particles with either definite momenta or definite positions.

Within quantum field theories, the concept of a particle with a definite position is not quite well understood, and there are confusing ideas about that notion. I will show that once this concept is properly formulated, quantum field theories can be employed in creating continuous sets of particles at different positions. In particular, such sets can be strings or branes. The quantization of a generic brane has so far been very difficult and has not yet been completely finished. However, with branes generated by means of the creation operators of a quantum field theory, we have solved the problem of the quantum brane. Now, if we consider the braneworld scenario, which regards our universe as a brane embedded in a higher-dimensional space, with the effective induced metric on the brane being the spacetime metric, we arrive at quantum gravity.

Acknowledgement

The book is a result of my long term research which was inspired in the contacts with many people. Special thanks I owe to Erasmo Recami, Asim O. Barut, Pierro Caldirola, Waldyr Rodrigues, Jr., Milutin Blagojecić, Larry Horwitz, and John Fanchi. Very useful and insightful were discussions with Victor Tapia, Marcos Maia, Martin Land, Igor Kanatchikov, Carlos Castro, Antonio Aurilia, William Pezzaglia, Aharon Davidson, and Daniel Grumiller. Each of those physicists improved my understanding of theoretical physics and helped me to proceed further. The support of the Jožef Stefan Institute and Slovenian Ministry of Science is greatly acknowledged. Last, but not least, I wish to express my thanks and gratitude to Mojca Vizjak-Pavšič, whose love and support was invaluable for me.

Acknowledgement

[illegible faded text]

Contents

Chapter 1

About Historical Misconceptions in Physics

*All truth passes through three stages. First, it is
ridiculed. Second, it is violently opposed. Third,
it is accepted as being self-evident.*

Arthur Schopenhauer

Dissemination of a truth, novel insight, or discovery is not always straightforward or easy. Just the contrary. In this respect, human society does
not behave much better than the groups of monkeys in numerous experiments[1]. For instance, the experimenters trained a low-rank-status monkey
in isolation in how to obtain food from a box or similar equipment. When
later observed by other monkeys, they very likely did not imitate this individual. He could be retrieving bananas from the automaton in front of the
eyes of other monkeys, but no one took notice and imitated him. However,
when the experimenters taught a high-rank-status monkey to retrieve food
from the machine, the other monkeys started to imitate him and obtain
bananas. Many other experiments show that chimpanzees, when given the
opportunity to watch alternative solutions to a foraging problem, prefer-
ably copy the method shown by the older, higher-ranking individual with
a prior trackrecord of success [2]. Records in the wild showed that majority of chimpanzee innovations are performed by low-ranking individuals[2],
and consequently, their spreading within the community is impeded, because not observed and imitated by others. Such studies suggest that most
chimpanzee innovations spread with difficulty, if ever.

---

[1]See, e.g., Refs. [2,3], and references therein.
[2]Most likely this is a means of circumventing competition from dominant group
members.

History teaches us that something analogous has been happening in human society all the time. The same story has repeated itself in numerous cases, from small inventions, such as the umbrella, to the large scientific discoveries, including the rotation of the Earth, molecular origin of heat, continental drift, quasicrystals, etc. In all those cases, there are only one or a few persons with insight, while a vast majority of people ignore, ridicule, or strongly oppose them. In the following, I will review some examples in which a discovery was rejected because of the misconceptions on the part of the scientific community of the epoch, or even worse, just ridiculed without any, though wrong, argumentation. Then, I will note that, in this respect, history has not had the final word and that the situation is very likely being repeated today and will be repeated in the future. Throughout this book, I will discuss a number of cases in which, in my opinion, the scientific community was too quick in rejecting a topic as physically unviable.

Rotation of the Earth

At the time of Galileo, scientists had difficulties in accepting the idea that the Earth was rotating. They raised various objections that supposedly falsified the Earth's motion. Namely, when one jumps, the flour would be displaced when one lands again. Similarly, when one drops a stone from a tower. Cannons would fire farther towards the west than towards the east. The birds could not fly because of the strong wind of thousands kilometers per hour. These and similar objections were nicely described in Galileo's "Dialogs" [4]. Such objections made sense within the Aristotelian physics in which every motion was caused by a force, and the concept of inertial motion, for which no force was needed, did not exist. Aristotel's physics predicted a number of phenomena that would manifest themselves if the Earth rotated. None of such phenomena were observed and the conclusion was that the Earth did not rotate.

A change of paradigm renders the situation quite different. If one accepts Galileo's principle of inertia, according to which, in the absence of force, there exists inertial motion, then it becomes evident that we observe no such phenomena identified by many scientists who were thinking within the paradigm of Aristotle's physics.

Influence of the Moon on the tides

A different kind of misconception concerning the observed phenomena being in contradiction with the accepted paradigm is the following.

Fishermen routinely observed the connection between the moon and the tides. Such a connection was considered by academicians to be a fiction of fishermen and denied any possibility of the far Moon influencing the sea.

Atomic theory and Statistical mechanics

Ludwig Boltzmann developed the kinetic theory, which postulated the existence of atoms and molecules. He found the connection between entropy S and probability W expressed by his famous formula $S = k_B \ln W$, where k_B is the constant bearing his name. At that time Boltzmann's theory was not generally appreciated. Among others, he had problems with the editor of an important German physics journal, who requested Boltzmann to refer to atoms and molecules only as conventional theoretical constructs but not as really existing objects. After years of fighting for the atom theory to be accepted, Boltzmann ran into mental health problems and committed suicide. This was only a few years before the new evidence convinced the world that atoms actually exist.

The problem of 3/4 in the electron mass

In more recent times there was the problem that the energy of the electric self-field of a moving charged object, such as the electron, was not the same as its inertial mass. Their ratio was calculated to be 3/4. This was the notorious problem of 3/4 for the electron mass. One obtains such a result if one calculates the 4-momentum of the electromagnetic field around a moving object by integrating the stress-energy tensor $T_{\mu\nu}$ over the wrong hypersurface. Many authors when calculating the electromagnetic energy of the electron at rest took the hypersurface element $d\Sigma = (d\Sigma_0, 0, 0, 0)$, where $d\Sigma_0 = d^3\boldsymbol{x} = dx^1 dx^2 dx^3$. Assuming that the electron radius is r_e, they obtained

$$P_{\mathrm{EM}}^0 = \int d^3\boldsymbol{x}\, T_{\mathrm{EM}}^{00} = \frac{e^2}{2r_e}. \tag{1.1}$$

Then, they calculated the momentum according to

$$P_{\mathrm{EM}}^i = \int d^3\boldsymbol{x}\, T_{\mathrm{EM}}^{i0} = \frac{2}{3}\frac{e^2}{r_e}u^i = m_e u^i\,, \quad i = 1, 2, 3. \tag{1.2}$$

The ratio was thus $P_{\mathrm{EM}}^0/m_e = 3/4$, which was a substantional puzzle, as also mentioned by Feynman in his Lectures on Physics [5].

Curiously, as pointed out by Rohrlich's review "The Saga of 3/4" [6], the correct solution has been obtained several times in history, also by Fermi, but repeatedly ignored or dismissed by the rest of the physics community. And what is the correct calculation of the momentum associated with the electromagnetic field of a moving electron or any other charged particle? Very straightforward: the correct covariant expression is

$$P_{\text{EM}}^\mu = \int \mathrm{d}\Sigma_\nu \, T_{\text{EM}}^{\mu\nu}. \tag{1.3}$$

If the object is at rest, then the latter expression becomes (1.1). However, if the object is moving, then the stress-energy tensor, of course, is no longer the same: it is Lorentz transformed. But Lorentz transformed is also the hypersurface $\mathrm{d}\Sigma_\nu$. It is no longer $\mathrm{d}\Sigma_\nu = (\mathrm{d}\Sigma_0, 0, 0, 0)$, but $\mathrm{d}\Sigma_\nu = (\mathrm{d}\Sigma_0, \mathrm{d}\Sigma_1, \mathrm{d}\Sigma_2, \mathrm{d}\Sigma_3)$. Therefore, instread of (1.2) one must take

$$P_{\text{EM}}^i = \int \mathrm{d}\Sigma_\nu \, T_{\text{EM}}^{i\nu}, \tag{1.4}$$

which brings the additional terms so that the result is

$$P_{\text{EM}}^i = \frac{e^2}{2r_e} u^i = m_e u^i. \tag{1.5}$$

This means that the correct calculation gives $m_e = P_{\text{EM}}^0$, as it should be.

In other words, if one calculates the electron's electromagnetic mass m_e in its rest frame as a null component of the 4-momentum, and then performs the calculation of P_{EM} in the frame in which the electron is moving, then one has to take the appropriately transformed hypersurface element $\mathrm{d}\Sigma_\nu$. The fact that so many authors did not take care of taking the properly transformed $\mathrm{d}\Sigma_\nu$ has roots in the fact that the field around a moving object satisfied $T_{\text{EM},\nu}^{\mu\nu} = 0$. If so, they argued, then it does not matter over which space like hypersurface one performs the integration when calculating the integral (1.4). Namely,

$$\int \mathrm{d}^4 x \, T_{\text{EM},\nu}^{\mu\nu} = \oint \mathrm{d}\Sigma_\nu T_{\text{EM}}^{\mu\nu}$$

$$= \int_{\Sigma_2} \mathrm{d}\Sigma_\nu T_{\text{EM}}^{\mu\nu} - \int_{\Sigma_1} \mathrm{d}\Sigma_\nu T_{\text{EM}}^{\mu\nu} = P_{\text{EM}}(\Sigma_2) - P_{\text{EM}}(\Sigma_1) = 0. \tag{1.6}$$

The quantity $P_{\text{EM}}(\Sigma) = \int_\Sigma \mathrm{d}\Sigma_\nu T_{\text{EM}}^{\mu\nu}$ is conserved and is the same regardless of the chosen Σ. However, if we consider the electromagnetic field around the electron, then the total stres-energy tensor also contains the contribution of the the particle itself. Therefore $T_{\text{EM},\nu}^{\mu\nu}$ is not zero at the position of the electron. If in the electron's rest frame Σ is oriented so that the normal to Σ points into the direction of the electron's worldline, the same arrangement must be taken in any other reference frame. Therefore, in the reference frame with respect to which the electron moves, the surface Σ should not be taken as in Eq. (1.2), i.e., not being oriented along x^0; it must be oriented along the particle's 4-velocity $u^\mu = (u^0, u^i)$.

Rocket drive

An American rocketry pioneer, Robert H. Goddard was the first man to make a multistage rocket and the first to make a liquid-fueled rocket. At that time it was believed that a rocket jet needs to be pushed against the air and that a rocket cannot fly in a vacuum. A paper in which Goddard presented in detail the principle of rocket propulsion was noticed by the press and subjected to violent criticism and ridicule. A culmination was on January 13, 1920, when the New York Times published an editorial entitled "A Severe Strain on Credulity" in which an unsigned "expert" stated

> That Professor Goddard, with his "chair" in Clark College and the countenancing of the Smithsonian Institution, does not know the relation of action to reaction, and of the need to have something better than a vacuum against which to react–to say that would be absurd. Of course, he only seems to lack the knowledge ladled out daily in high schools[3].

This article damaged Goddard's career, but he continued his work with rocket experiments. He died in 1945, when the rockets of a similar but much improved design, were used by Germany as weapons. His dreams of reaching the Moon were fulfilled only 24 years later.

Continental drift

The idea that the Americas and Africa were once joined together into a single continental mass and later torn apart has been held by numerous scientists, including geologists. However, it was Alfred Wegener who proposed a more complete theory in 1912. But it was not accepted because the evidence provided by Wegener was insufficient, especially because of the lack of a mechanism for continental drift. Moreover, Wegener's estimate for the speed of continental motion, namely, 250 cm/year, was much too high. And yet, later it turned out that Wegener was correct and that the continents indeed move, although at a speed a hundred times slower. The mechanism

---

[3] "Topics of the Times". New York Times. January 13, 1920. In July 1969, when Apollo 11 landed on the Moon, the New York Times published a short "Correction", stating that "... a rocket can function in a vacuum as well as in an atmosphere".

causing continental drift was found to be the convection currents beneath the Earth's crust. Wegener, who was not geologist, proposed a theory that had several flaws, but his general idea was correct. This story is instructive because it teaches us that, first, a breakthrough insight need not come from a professional and, second, even if certain details of a proposed theory are incorrect, the proposal as a whole might have value for advancing a scientific discipline. There are numerous other similar examples in the history of science. Therefore, if the protagonist is an amateur, (s)he should not be automatically dismissed, and flaws in the theory could later be eliminated.

Quarks

After defending his PhD thesis, George Zweig proposed the theory of quarks at the same time as the renown theoretical physicist Murray Gell-Mann. They submitted their papers to the same journal which then published Gell-Mann's and rejected Zweig's. Five years later Gell-Mann received a Nobel Prize in Physics for his work on particle physics but not explicitly for the quark theory, because at that time it was not yet generally accepted.

What happened to young Zweig, concerning the recognition of his discovery by the journal, has a parallel with what has been observed in chimpanzee's communities, whose members paid attention and copied the inventions of higher-ranking individuals but ignored those found by young or low ranking individuals. However, regardless of Gell-Mann's and Zweig's "ranks", their theory of quarks as real particles did not receive wide recognition at that time and was even not mentioned by the Nobel committee in the justification of the prize.

Quasicrystals

One could say that those stories belong to the past and that nowadays something similar is no longer happening. Is that indeed so? In 1982 Dan Shechtman discovered quasicrystals. What then happened is nicely explained in the Wikipedia article[4]:

> From the day Shechtman published his findings on quasicrystals in 1984 to the day Linus Pauling died (1994), Shechtman experienced hostility from him toward the non-periodic interpretation. "For a long time it was me against the world," he said. "I was a subject of ridicule and lectures about the basics

---

[4]https://en.wikipedia.org/wiki/Dan_Shechtman

of crystallography. The leader of the opposition to my findings was the two-time Nobel Laureate Linus Pauling, the idol of the American Chemical Society and one of the most famous scientists in the world. For years, 'til his last day, he fought against quasi-periodicity in crystals. He was wrong, and after a while, I enjoyed every moment of this scientific battle, knowing that he was wrong." [16]

Linus Pauling is noted saying "There is no such thing as quasicrystals, only quasi-scientists." [17] Pauling was apparently unaware of a paper in 1981 by H. Kleinert and K. Maki which had pointed out the possibility of a non-periodic Icosahedral Phase in quasicrystals[18] (see the historical notes). The head of Shechtman's research group told him to "go back and read the textbook" and a couple of days later times "asked him to leave for 'bringing disgrace' on the team." [19] Shechtman felt dejected.[17] On publication of his paper, other scientists began to confirm and accept empirical findings of the existence of quasicrystals.[20][21]5

In 2011, Shechtman was awarded the Nobel prize in Chemistry. "Once-Ridiculed Discovery Redefined the Term Crystal" was the title of an article in Science (vol. 334, 14 October 2011), which started with the words: "Daniel Shechtman has the last laugh."

In all the cases described above, and many other cases, the vast majority of the scientific community was unable or unwilling to recognize the importance of a discovery. Instead, in the best case, if not ignoring or ridiculing the discoverer, scientists raised various arguments that purportedly proved the incorrectness of the idea or the theory. Sometimes, as was the

---

5 Quoted references:

16. "Iowa State, Ames Laboratory, Technion scientist wins Nobel Prize in Chemistry" (Press release). Ames, Iowa: Iowa State University. 2011-10-05.

17. Lannin, Patrick (2011-10-05). "Ridiculed crystal work wins Nobel for Israeli". Reuters. Retrieved 2011-10-22.

18. Kleinert H., Maki K. (1981). "Lattice Textures in Cholesteric Liquid Crystals" (PDF). Fortschritte der Physik. 29 (5): 219–259. Bibcode:1981ForPh..29..219K. doi:10.1002/prop.19810290503

19. Jha, Alok (5 Jan 2013). "Dan Shechtman: 'Linus Pauling said I was talking nonsense'". Guardian.

20. Bradley, David (Oct 5, 2011). "Dan Shechtman discusses quasicrystals". Science-Base. Retrieved 5 October 2011. Shechtman video interview

21. "Clear as crystal". Haaretz. 2011-04-01. Retrieved 2011-10-06.

case with quasicrystals, they even rejected and ridiculed the experiment, if it contradicted the existing, widely accepted theory[6].

After all those shocking cases, how can we indeed be sure that something similar is not happening right now? Have people indeed changed so much that they are nowadays no longer apt to behave in the way they used to in the past? I mean, of course, they still do behave more or less similarly in all respects. Why then should they have a different attitude against the claimed "out of the box" discoveries and their protagonists? The vast majority certainly does not behave differently. Today people ridicule or consider many topics to be "pathological science". But how can these individuals be so sure that there are no new quasicrystals among the ridiculed and rejected topics and that they do not behave as Shechtman's opponents did? After all, the arguments based on the theoretical impossibility of crystals with fivefold symmetry seemed to be very sound.

One could say "Yes, but there are so many people advocating "genial" unusual ideas and strange theories that are inconsistent with the accepted knowledge. All such ideas are likely wrong and without any value. Serious scientists should not waste their precious time considering such lunatic subjects." How to recognize among so many obviously wrong ideas the right one? If the suggested idea contradicts textbooks, then people usually think that it must be wrong, and that there is no need to give a second thought. The safest attitude is to ignore, reject, and ridicule it for fun, or in the best case, raise seemingly impeccable counterarguments. The result of such an approach is that people miss an important new discovery, as has happened throughout history and is certainly also happening today.

By analogy, a bad or average doctor overlooks a patient who has a very rare disease, but most of the symptoms indicate a banal disease. Such a doctor then does not prescribe adequate treatment, and consequently, in the worst case, the patient dies. When a good doctor, despite facing every day patients who have usual diseases, encounters a patient with a rare but dangerous disease, (s)he takes care, recognizes among the usual symptoms also the unusual ones, prescribes additional examinations, and makes the correct diagnosis. Similarly, a good scientist recognizes a brilliant idea among many seemingly crackpot ideas. Such cases have been repeatedly happening throughout the history of science in general, and physics in particular,

---

[6]Such cases disprove the school and textbook simplistic wisdom that a theory is falsified by an experiment. Often the actual course of events is that if an experiment disagrees with the existing theory described in textbooks, not the theory but the experiment is rejected, and the experimenter ridiculed and dismissed.

and they are certainly happening now as well. This way science makes a paradigm shift despite the opposition of the vast majority of scientists, who are unable or unwilling to step out of the boxes to which their thoughts are constrained. Some topics that today are considered problematic and physically unviable, rejected by the argumentation that I perceive as flawed or resting on misconceptions, will be considered throughout this book.

Chapter 2

Higher Derivative Theories and Negative Energies

A great stumbling block in current theoretical physics is the concept of negative energies. When they first occurred in relativistic mechanics and subsequently in the Dirac equation, there was initially much puzzlement among physicists. Problematic was the issue of stability of such systems. So Fermi [8] wrote:

> It was well known that the most serious difficulty in Dirac's relativistic wave equation lies in the fact that it yields besides the normal positive states also negative ones, which have no physical significance. This would do no harm if no transition between positive and negative states were possible (as are, e.g., transitions between states with symmetrical and antisymmetrical wave function). But this is unfortunately not the case: Klein has shown by very simple example that electrons impinging against a very high potential barrier have a finite probability of going in a negative state.

Fortunately, Dirac found an elegant solution by assuming the existence of a "sea" filled with negative energy electrons. Because electrons are fermions, every state can be occupied by only one electron, and they can indeed fill such a sea. Although the concept of the Dirac sea has become obsolete with the development of quantum field theories, it has recently been revitalized within the approaches in which fermionic quantum fields are elements of infinite-dimensional Clifford algebras [9].

Although the negative energy states in the Dirac equation were first considered problematic (today some would say "pathological"), physicists of that time did not reject the Dirac equation. Instead, to the benefit of physics, they insisted on finding a solution to the puzzle, and they did find it. Today, in the attempts to quantize gravity, we have again encountered negative energies. Namely, while ordinary gravity with the Einstein-Hilbert action, containing the curvature scalar R, is not renormalizable, it was found that a higher order gravity, such as $R + R^2$, is renormalizable. Unfortunately, such theories contain higher derivatives, and due the seminal work by Ostrogradski [10], are generally believed to be unstable. The so called "Ostrogradski instability" has become one of the biggest stumbling blocks against quantum gravity.

2.1   Negative energies and stability

Newtonian dynamics and most of the dynamics usually considered in physics, including special and general relativity, is based on the Lagrangians that contain first order derivatives with respect to time. A typical action has the form

$$I = \int dt\, L(x^\alpha, \dot{x}^\alpha), \tag{2.1}$$

from which one derives the Euler-Lagrange equations of motion for the variables x^α:

$$\frac{d}{dt}\frac{\partial L}{\partial \dot{x}^\alpha} - \frac{\partial L}{\partial x^\alpha} = 0. \tag{2.2}$$

For instance, a free non relativistic particle satisfies the action principle

$$I = \int dt \frac{1}{2} m \dot{x}^\alpha \dot{x}^\beta \delta_{\alpha\beta}, \tag{2.3}$$

with the corresponding equations of motion

$$\ddot{x}^\alpha = 0, \tag{2.4}$$

whose general solution is

$$x^\alpha = x^\alpha(0) + \dot{x}^\alpha(0)t. \tag{2.5}$$

For a chosen initial conditions $x^\alpha(0)$, $\dot{x}^\alpha(0)$, this describes a straight trajectory with the uniform velocity.

A more general Lagrangian is

$$L = \frac{m \dot{x}^\alpha \dot{x}^\beta \delta_{\alpha\beta}}{2} - V(x), \tag{2.6}$$

which gives the equations of motion

$$m\ddot{x}^\alpha + \frac{\partial V}{\partial x^\alpha} = 0. \tag{2.7}$$

As a model potential one often considers that of a harmonic oscillator,

$$V = \frac{1}{2} k x^\alpha x^\beta \delta_{\alpha\beta}, \tag{2.8}$$

in which case the equations of motion are

$$m\ddot{x}^\alpha + kx^\alpha = 0. \tag{2.9}$$

This describes an oscillatory motion with the frequancy $\omega = \sqrt{k/m}$ if $k > 0$. This system exhibits a stable oscillatory motion if the potential is bounded from below, and an unstable, run away motion, if the potential is unbounded from below.

However, if instead of the Lagrangian (2.6) we take the Lagrangian

$$L = -\frac{m\dot{x}^\alpha \dot{x}^\beta \delta_{\alpha\beta}}{2} + V(x), \tag{2.10}$$

then the equations of motion remain the same, i.e., Eq. (2.7). But the criterion for stability is then reversed. Namely, the potential with a minimum now gives instability, whilst the potential with a maximum gives stability of the system described by the Lagrangian (2.10).

Having clarified that, let us now consider the following Lagrangian

$$L = \frac{1}{2}m\dot{x}^\alpha \dot{x}^\beta \eta_{\alpha\beta} - V(x) \tag{2.11}$$

where $\eta_{\alpha\beta} = \text{diag}(1,1,1,...,-1,-1,-1,...)$ is the metric with a pseudo euclidean signature (p,q) i.e., $(++...---)$, with p plus and q minus signs. The kinetic energy of the system, $E_{\text{kin}} = \frac{1}{2}m\dot{x}^\alpha \dot{x}^\beta \eta_{\alpha\beta}$, is then the sum of the components with positive and negative values.

The equations of motion derived from the Lagrangian (2.11) are

$$m\ddot{x}^\alpha + \eta^{\alpha\beta}\frac{\partial V}{\partial x^\beta} = 0. \tag{2.12}$$

The force $F^\alpha = -\eta^{\alpha\beta}\frac{\partial V}{\partial x^\beta}$ does not depend only on the potential, but also on the metric $\eta^{\alpha\beta}$. The degrees of freedom x^α, $\alpha = 1,2,...,p$, with positive kinetic energies experience the force $F^\alpha = -\frac{\partial V}{\partial x^\alpha}$, whilst the degrees of freedom x^α, $\alpha = 1,2,...,q$, with negative kinetic energies experience the force $F^\alpha = \frac{\partial V}{\partial x^\alpha}$. Therefore, while the criterion for stability of the positive energy degrees of freedom is *minimum* of a potential, the criterion for stability of negetaive energy degrees of freedom is *maximum* of the potential.

A degree of freedom x^α is thus stable at $x^\alpha = x_0^\alpha$ if

$$\frac{\partial V}{\partial x^\alpha} = 0 \,, \qquad \eta^{\alpha\beta}\frac{\partial^2 V}{\partial x^\alpha \partial x^\beta}\bigg|_{x_0^\alpha} > 0 \quad \text{(no sum)}, \tag{2.13}$$

and if $\frac{\partial^2 V}{\partial X^\alpha \partial x^\beta} = 0$ for all values of $x^\beta \neq x^\alpha$, i.e., if there are no coupling terms between x^α and x^β, $\beta \neq \alpha$.

For instance, if there are only two degrees of freedom, $x^\alpha = (x, y)$, the
the Lagrangian that contains negative kinetic energy is

$$L = \frac{m}{2}(\dot{x}^2 - \dot{y}^2) - V(x, y), \tag{2.14}$$

and the equations of motion are

$$m\ddot{x} + \frac{\partial V}{\partial x} = 0, \tag{2.15}$$

$$m\ddot{y} - \frac{\partial V}{\partial y} = 0. \tag{2.16}$$

Then Eq. (2.13) reads

$$\frac{\partial V}{\partial x}\Big|_{x_0} = 0 \ , \qquad \frac{\partial^2 V}{\partial x^2}\Big|_{x_0} > 0, \tag{2.17}$$

$$\frac{\partial V}{\partial y}\Big|_{y_0} = 0 \ , \qquad \frac{\partial^2 V}{\partial y^2}\Big|_{x_0} < 0, \tag{2.18}$$

and

$$\frac{\partial^2 V}{\partial x \partial y} = 0. \tag{2.19}$$

Taking as an example $V(x, y) = \frac{k}{2}(x^2 - y^2)$, we see that the above criterion
for stability is indeed satisfied. The equations of motion are the same for
x and y.

Generalizing this to any dimensions, we have

$$L = \frac{1}{2}m\dot{x}^\alpha \dot{x}^\beta \eta_{\alpha\beta} - \frac{k}{2}x^\alpha x^\beta \eta_{\alpha\beta}. \tag{2.20}$$

The equations of motion are

$$m\ddot{x}^\alpha + kx^\alpha = 0. \tag{2.21}$$

These are all the same harmonic oscillator equations. We see that in the case
of a system of uncoupled harmonic oscillators it does not matter whether
a degree of freedom has positive or negative kinetic energy, provided that
also the corresponding term in the potential has appropriate sign.

I am explaining all this at length, because according to my experience,
a brief explanation was often not understood correctly. It was typically
followed by objections that a system with a potential unbounded from below
cannot be stable. People somehow overlooked the points discussed above,
because they were so much accustomed to thinking that a potential must
have a minimum (at least a local one), otherwise the system is unstable.

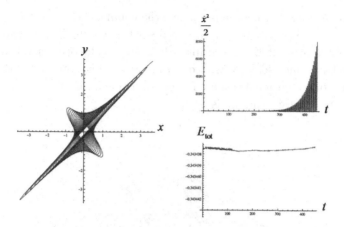

Fig. 2.1 The solution of the system described by eqs. (2.23),(2.24), with $m = 1$, $k = 1$, $\lambda = 0.1$, for the initial conditions $\dot{x}(0) = 1$, $\dot{y}(0) = -1.2$, $x(0) = 0$, $y(0) = 0.5$.

However, in the literature we can find a consensus among the informed experts, who are aware of the points discussed in this chapter, that the presence of the degrees of freedom with negative energies implies instability anyway. Namely, if there is coupling between positive and negative energy degrees of freedom, then the energy flows between them and causes a runaway behaviour of the system. We will show that such a runaway behaviour indeed happens—in the presence of physically unrealistic potentials.

Let us consider a system with two degrees of freedom, described by the Lagrangian (2.14). Suppose that the potential is

$$V = \frac{k}{2}(x^2 - y^2) - \frac{\lambda}{4}(x^2 - y^2)^2, \tag{2.22}$$

so that besides the terms of the harmonic oscillators it contains also an interaction term. Because $\frac{\partial^2 V}{\partial x \partial y} = -2\lambda xy \neq 0$, such potential couples x and y. The equations of motion are

$$m\ddot{x} + kx + \lambda x(x^2 - y^2) = 0, \tag{2.23}$$

$$m\ddot{y} + ky + \lambda y(x^2 - y^2) = 0. \tag{2.24}$$

Using Mathematica, we have found numerical solutions of this system which show that the trajectory in the (x, y)-space runs into infinity [11], although it started near the origin $x(0) = 0$, $y(0) = 0.5$ (Fig. 2.1, left), with the finite velocity $\dot{x}(0) = 1$, $\dot{y}(0) = -1.2$. The velocity and the kinetic energy for each component x and y also run into infinity (Fig. 2.1 right), whilst the total energy

$$E_{\text{tot}} = \frac{m}{2}(\dot{x}^2 - \dot{y}^2) + V(x, y) \tag{2.25}$$

remains constant.

Numerical calculations thus confirm the common wisdom that negative energies lead to instability. A system described by the Lagrangian (2.14) with the potential (2.22) is unstable. However, on the other hand we know that the harmonic oscillator with quadratically rising potential of the form of a parabola is an idealization. In reality a potential does not grow into infinity. A realistic potential is finite.

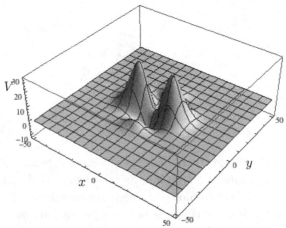

Fig. 2.2   An example of the potential that is bounded from below and from above (see Eq. (2.28)).

One obtains a quadratic potential as an approximation of the expansion of a symmetric potential

$$V(x) = V(x_0) + \frac{1}{2}(x - x_0)^2 \frac{\partial^2 V}{\partial x^2}|_{x=x_0} + \mathcal{O}(x^4). \tag{2.26}$$

The infinity of the remaining terms, denoted by $\mathcal{O}(x^4)$, is omitted. Similarly, in the case of two variables we have

$$V(x,y) = V(x_0, y_0) + \frac{1}{2}(x - x_0)^2 \frac{\partial^2 V}{\partial x^2}\bigg|_{x_0} + \frac{1}{2}(y - y_0)^2 \frac{\partial^2 V}{\partial y^2}\bigg|_{y_0}$$

$$+(x - x_0)(y - y_0) \frac{\partial^2 V}{\partial x \partial y}\bigg|_{x_0, y_0} + \frac{1}{4!}\left((x - x_0)^4 \frac{\partial^4 V}{\partial x^4}\bigg|_{x_0} + (y - y_0)^4 \frac{\partial^4 V}{\partial y^4}\bigg|_{y_0}\right.$$

$$\left.+(x - x_0)^2(y - y_0)^2 \frac{\partial^4 V}{\partial x^2 \partial y^2}\bigg|_{x_0, y_0}\right) + ... \tag{2.27}$$

The potential (2.22) can thus be considered as a part of the above expansion for $x_0 = 0$, $y_0 = 0$. Let us therefore take the reasonable assumption that unbounded potentials like (2.22) are idealizations, just few terms of the expansion of a realistic, bounded, potential. If so, let us investigate what happens with stability of a system in the presence of a potential bounded from below and from above.

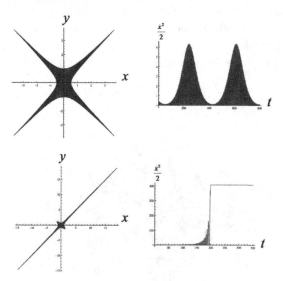

Fig. 2.3   The solution of the system described by eqs. (2.14),(2.28), with $m = 1$, $k = 1$, $\lambda = 0.1$, $a = 0.001$, for the initial conditions $\dot{x}(0) = 1$, $\dot{y}(0) = -1.2$, $x(0) = 0$, $y(0) = 0.5$ (up). The same system as in Fig. 2.3, except that the initial velocity is increased from $\dot{x}(0) = 1$ to $\dot{x}(0) = 1.5$ (down).

As an example let us consider the following potential (Fig. 2.2)

$$V = e^{-a(x^2+y^2)} \left[\frac{1}{2}(x^2 - y^2) + \frac{\lambda}{4}(x^2 - y^2)^2 \right]. \tag{2.28}$$

which is the potential (2.22) multiplied with an exponentially damped function. In Fig. 2.3 (up) there are results for numerical solutions of the equations of motion of the Lagrangian (2.14) for the same parameters and the initial data as in Fig. 2.1, and for the damping constant $a = 0.001$. Now the system, instead of escaping into infinity, oscillates within a finite region of the (x, y)-space.

If we increase the velocity, then the system, after oscillating for some time within a finite region, escapes into infinity, as shown in Fig. 2.3 (down). With time the amplitude of oscillations increases until the particle escapes towards infinity with a constant final velocity. Such system is therefore stable in the sense that its velocity does not increase without cease towards the infinity. We have thus a system with positive feedback, where the x and y degrees of freedom are feeding each other towards higher and higher amplitudes, until the process stops and reaches a final plato, as a consequence of the bounded potential. This is similar to the familiar positive feedback between a microphone and a loudspeaker, resulting in ever increasing intensity of the sound, which does not grow to infinity, because of the final power of the amplifier.

As another example let us consider the potential (Fig. 2.4)

$$V = \frac{k}{2}(\sin^2 x - \sin^2 y) + \lambda \sin x \sin y. \qquad (2.29)$$

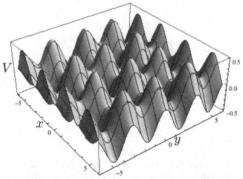

Fig. 2.4   Another example of the potential that is bounded from below and from above (see Eq. (2.29)).

The Lagrangian (2.14) then gives the following equations of motion

$$m\ddot{x} + k\sin x \cos x + \lambda\cos x \sin y = 0, \qquad (2.30)$$

$$m\ddot{y} + k\sin y \cos y - \lambda\cos y \sin x = 0. \qquad (2.31)$$

Numerical calculations reveal that such system in most cases behaves normally [12]. The x and y components of the velocity and kinetic energy remain finite. Depending on the initial velocity, the x and y either oscillate permanently (Fig. 2.5) within a valley or a hill of the potential (Fig. 2.4), or escape and wander around with velocity that never escapes into infinity as shown in Fig. 2.6. This was true for $k = 1$ and the coupling constant $\lambda = 0.1$.

If we increase the coupling constant to $\lambda = 0.3$, then for the same initial data as in Fig. 2.6 the system exhibits large jumps in the (x, y) and the velocity space (Fig. 2.7).

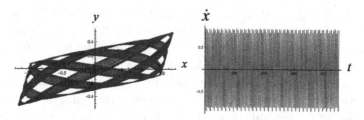

Fig. 2.5 Trajectory in the (x, y)-space (left) and velocity oscillations (right) for the system described by the Lagrangian (2.14) with the potential (2.29) ($\lambda = 0.1$, $\dot{x} = 0$, $\dot{y} = 0.2$, $x(0) = 1$, $y(0) = 0$).

Another possibility is to start the evolution at $x(0) = 0$, $y(0) = 0$, and with the initial velocity $\dot{x}(0) = 0.1$, $\dot{y}(0) = 0$. Then for a certain period from $t = 0$ to $t = t_c$ the potential is approximately $V = \frac{k}{2}(x^2 - y^2) + \lambda xy$, which in the case of the Lagrangian (2.14) gives an unstable solution in the form of an outgoing spiral (Fig. 2.8 left). At the time around $t = 23$ the trajectory deviates from the spiral and at $t = 80$ starts escaping far from the origin, with increasing velocity (Fig. 2.8 right, Fig. 2.9 left), which at $t = 998$ reaches a maximum value $\dot{x} = 89.4$ (Fig. 2.9 right), the position being $x = 52130$, $y = 52150$.

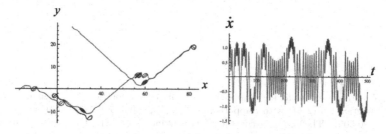

Fig. 2.6 Trajectory in the (x, y)-space and velocity oscillations for the system described by the Lagrangian (2.14) with the potential (2.29) ($\lambda = 0.1$, $\dot{x} = 1$, $\dot{y} = 0.9$, $x(0) = 2.5$, $y(0) = -0.505$). The initial velocity is now increased.

Such escape of the system with velocity starting to increase up to a maximum, much higher than the potential difference, takes place along a "diagonal" directions defined, e.g, by $x = y$. The system coordinates and velocity then oscillate around $x = y$ and $\dot{x} = \dot{y}$, and, because the y-component of the kinetic energy is negative, self-accelerates itself as shown in Fig. 2.9. Depending on the initial conditions, such "jumps" can be very high, perhaps even infinite. Namely, in a run in which the initial data were

$\dot{x}(0) = 0.001$, $\dot{y}(0) = 0$, a huge jump started at $t = 2900$. The velocity increased linearly, and at $t = 4000$, where $\dot{x}$ was 162.6, there was not yet a sign of a turn down. In yet another run the initial speed was $\dot{x}(0) = 10^{-6}$, and at $t = 4000$ it reached the value of 78.8, with no maximum at sight. This time the factor of velocity increase with respect to the initial velocity was 7.88×10^{10}. A warning has to be mentioned at this point that numerical errors over such longs time intervals, implying many steps, accumulate to such extent that the calculated functions $x(t)$, $y(t)$, $\dot{x}(t)$, and $\dot{y}(t)$, are not solution to our system over all the interval from $t = 0$ to $t = T$. However, they are solutions in the vicinity of any $t \in [0, T]$. And so we found that at certain $t = t_c$ the position and velocity are such that the system starts making a huge jump. This can happen if the wavy-like potential (such as that of Fig. 2.4) with well defined diagonal directions extents to infinity, which is again a nonrealistic potential.

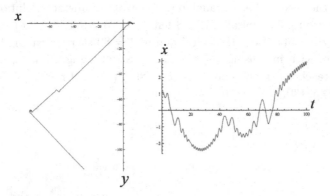

Fig. 2.7   Trajectory in the (x, y)-space and velocity oscillations for the system described by the Lagrangian (2.14) with the potential (2.29) ($\lambda = 0.3$, $\dot{x} = 1$, $\dot{y} = 0.9$, $x(0) = 2.5$, $y(0) = -0.505$). The coupling constant λ is now increased, whilst the initial data are the same as in Fig. 2.6.

In the presence of a damped potential such as

$$V = \left(\frac{k}{2} (\sin^2 x - \sin^2 y) + \lambda \sin x \sin y \right) \left(1 + \tanh(r_0 - \sqrt{x^2 + y^2}) \right) \tag{2.32}$$

which vanishes outside a domain defined by r_0 (see Fig. 2.10 up), the system's velocity stabilizes to a constant value. An example is shown in Fig. 2.10.

Next example is the scattering of the particle. satisfying the equations of motion derived from the Lagrangian (2.14) in the presence of the potential

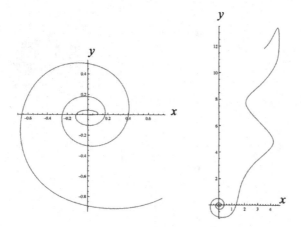

Fig. 2.8 Trajectory in the (x, y)-space for the system described by the Lagrangian (2.14) with the potential (2.29) ($\lambda = 0.3$, $\dot{x} = 0.1$, $\dot{y} = 0$, $x(0) = 0$, $y(0) = 0$). The system now starts from the coordinate origin, where $\sin x \approx x$ and $\sin y \approx y$ In the left plot the time intervals is from $t = 0$ to $t = 20$. In the right plot the time interval is extended to $t = 40$.

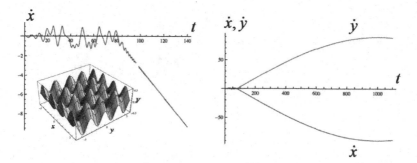

Fig. 2.9 Velocity $\dot{x}$ as function of time for the same initial data as in Fig.2.8. In the left plot the time interval is from $t = 0$ to $t = 140$. In the right plot the time interval is extended to $t = 2100$, and the upper curve belongs to $\dot{y}$.

(2.28), (Fig. 2.2) or the potential (2.32), (Fig. 2.10 up). In the considered examples, shown in Fig. 2.11, the initial velocities are small, whilst the final velocities are high. The factors of increase are 13.8 and 23.8, respectively.

In all those cases with bounded potential a runaway behavior of the system's velocity does not happen. Velocity and hence the corresponding component of the kinetic energy remains finite. If, during the evolution of the system, there is an increase of the absolute value of the kinetic energy, it cannot exceed the overall potential difference[1]. Namely, the total energy

---

[1]This holds if the potential is asymptotically constant.

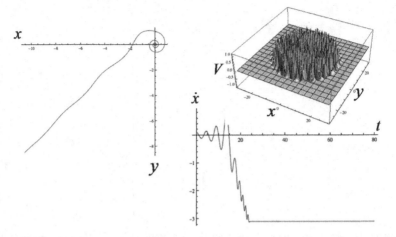

Fig. 2.10   Potential cut at $r_0 = 20$ ($\lambda = 0.3$, $\dot{x} = 0.1$, $\dot{y} = 0$, $x(0) = 0.0000001$, $y(0) = 0$). The system escapes with a constant velocity

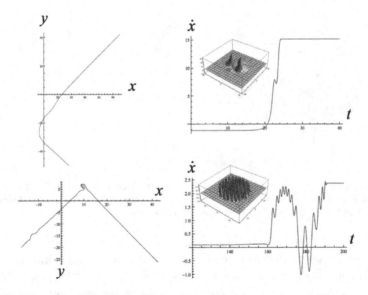

Fig. 2.11   Examples of the scattering on a potential giving a non vanishing force within a bounded region of the (x, y)-space, and zero force outside that region. a) Potential defined in Eq. (2.28), $\dot{x}(0) = -1.1$, $\dot{y}(0) = 1$, $x(0) = 28$, $y(0) = -50$ (up). b) Potential defined in Eq. (2.32), $\dot{x}(0) = 0.1$, $\dot{y}(0) = 0.10001$, $x(0) = -30$, $y(0) = -32$ (down).

is conserved, it is the same at all times,

$$E_x + E_y + V = C, \tag{2.33}$$

where C is a finite constant. Therefore,

$$E_x(t) + E_y(t) = E_x(0) + E_y(0) + V(t) - V(0), \tag{2.34}$$

which in the case of a bounded potential implies that $E_x(t) + E_y(t)$ is finite if $E_x(0) + E_y(0)$ is finite. Here $E_x = \dot{x}^2/2$ and $E_y = -\dot{y}^2/2$, and they typically oscillate out of phase. A consequence is that the positive term $E_x(t)$ and the negative term $E_y(t)$ cannot separately run away towards infinity, with their sum remaining finite. Each term, $E_x(t)$ and $E_y(t)$, remain finite. All our numerical caclulations confirm this.

However, there is a peculiar feature of such system containing positive and negative kinetic energies. Namely, the initial kinetic energies $E_x(0)$ and $-E_y(0)$ can be small, and then during the evolution of the system increase to the final values $E_x(t)$, $-E_y(t)$ which are significantly higher than $E_x(0)$, $-E_y(0)$. In the examples considered in our calculations the factor of increase could be for given initial conditions even around 100. Such increase of the kinetic energy is consistent with the energy conservation law (2.34). Because the x and y degrees of freedom are coupled, the energy flows between them. Thus the energy pulse from the y can push the x to slightly increase the amplitude of its oscillations, and vice versa. Such process can continue until the limit imposed by the absolute height of the potential V is reached. Then the run away behavior stops, and the system stabilizes itself. Of course, if the absolute height of the potential difference is infinite, then the runaway behaviour may not stop, but continue for ever. But such an unbounded potential is physically unrealistic.

2.2 Unequal masses, unequal tension: Special cases of generic metric

We have seen that stability of a system can be realized if the potential is bounded. Now we will show that in some cases stability can be as well achieved in the presence of an unbounded potential, such as (2.22), if masses, i.e., the constants multiplying each component of the acceleration, are different for different degrees of freedom. Then, instead of the equations of motion (2.23), (2.24), we have

$$m_x \ddot{x} + kx + \lambda x(x^2 - y^2) = 0, \tag{2.35}$$

$$m_y \ddot{y} + ky + \lambda y(x^2 - y^2) = 0. \tag{2.36}$$

Numerical calculations then show that such system is no longer unstable (Fig. 2.12). As shown in Refs. [11, 12], it exhibits stable behavior. Typical plots are very similar to those shown in Fig. 2.3 (up). A theoretical explanation of why this is so still needs to be found. A possible answer for a special case is given in Ref. [12].

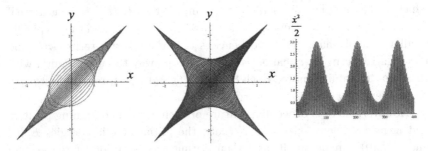

Fig. 2.12 Left and middle: The (x, y) plot of the solution to Eqs. (2.35), (2.36) for constants $m_x = 1/1.01$ and $m_y = 1$. Right: The kinetic energy $\dot{x}^2/2$ as function of time. The coupling constant is $\lambda = 0.1$.

Equations (2.35), (2.36) belong to the class of equations of motion derived from the Lagrangian

$$L = \frac{1}{2} g_{\alpha\beta} \dot{x}^\alpha \dot{x}^\beta - \frac{1}{2} k_{\alpha\beta} x^\alpha x^\beta - \frac{\lambda}{4} \lambda_{\alpha\beta\gamma\delta} x^\alpha x^\beta x^\gamma x^\delta \qquad (2.37)$$

for a particular choice of $g_{\alpha\beta}$, $k_{\alpha\beta}$ and $\lambda_{\alpha\beta\gamma\delta}$, which gives

$$L = \frac{1}{2} \left(m_x \dot{x}^2 - m_y \dot{y}^2 \right) - \frac{k}{2} (x^2 - y^2) - \frac{\lambda}{4} (x^2 - y^2)^2, \qquad (2.38)$$

with $m_x \neq m_y$ $k_x = k_y = k$.

Another possibility is such a choice of $g_{\alpha\beta}$, $k_{\alpha\beta}$ and $\lambda_{\alpha\beta\gamma\delta}$ which gives

$$L = \frac{1}{2} \left(\dot{x}^2 - \dot{y}^2 \right) - \frac{1}{2} (\omega_x^2 x^2 - \omega_y y^2) - \frac{\lambda}{4} (x + y)^4, \qquad (2.39)$$

with $m_x = m_y = 1$, $k_x \neq k_y$, and an appropriate choice of the coefficients $\lambda_{\alpha\beta\gamma\delta}$, i.e., $\lambda_{xxxx} = \lambda$, $\lambda_{xxxy} = \lambda_{xxyx} = \lambda_{xyxx} = \lambda_{yxxx} = \lambda/4$, etc.: We now write $k_x = \omega_1^2$, $k_y = \omega_2^2$.

Introducing the new variables

$$u = \frac{x + y}{\sqrt{2}}, \qquad v = \frac{x - y}{\sqrt{2}}, \qquad (2.40)$$

the Lagrangian (2.39) becomes

$$L = \dot{u}\dot{v} - \frac{1}{4} \left[(\omega_1^2 - \omega_2^2)(u^2 + v^2) + 2(\omega_1^2 + \omega_2^2)uv \right] - \lambda u^4. \qquad (2.41)$$

The equations of motion are

$$\ddot{u} + \frac{1}{2}(\omega_1^2 + \omega_2^2)u + \frac{1}{2}(\omega_1^2 - \omega_2^2)v = 0, \tag{2.42}$$

$$\ddot{v} + \frac{1}{2}(\omega_1^2 + \omega_2^2)v + \frac{1}{2}(\omega_1^2 - \omega_2^2)u + 4\lambda u^3 = 0. \tag{2.43}$$

Eliminating v from the first equation and inserting it into the second equation, we obtain

$$u^{(4)} + (\omega_1^2 + \omega_2^2)\ddot{u} + \omega_1^2\omega_2^2 u - \Lambda u^3 = 0, \tag{2.44}$$

where $\Lambda = 2(\omega_1^2 - \omega_2^2)\lambda$. We have thus obtained the forth order equation $(u^{(4)} \equiv \ddddot{u})$, known as the self-interactin Pais-Uhlenbeck oscillator, which can be written as

$$\left(\frac{d^2}{dt^2} + \omega_1^2\right)\left(\frac{d^2}{dt^2} + \omega_2^2\right)u - \Lambda u^3 = 0. \tag{2.45}$$

The corresponding Lagrangian is

$$L = \frac{1}{2}\left[\ddot{u}^2 - (\omega_1^2 + \omega_2^2)\dot{u}^2 + \omega_1^2\omega_2^2 u^2\right] + \frac{\Lambda}{4}u^4. \tag{2.46}$$

The action is given by the integral of the above Lagrangian over time, and after omitting the surface term, we have

$$I = \int dt \left[\frac{1}{2}u\left(\frac{d^2}{dt^2} + \omega_1^2\right)\left(\frac{d^2}{dt^2} + \omega_2^2\right)u + \frac{\Lambda}{4}u^4\right] \tag{2.47}$$

Pais-Uhlenbeck oscillator is a toy model for higher derivative field theories of the form

$$I = \int d^4x \left[\frac{1}{2}\phi\left(\Box + m_1^2\right)\left(\Box + m_2^2\right)\phi + \frac{\Lambda}{4}\phi^4\right] \tag{2.48}$$

in which the spatial dependence of the field $\phi(x)$, $x \equiv x^\mu = (t, \boldsymbol{x})$, is supressed. Higher derivative theories, including $R + R^2$ gravity, are usually considered as problematic, because they include negative energies.

We have seen that negative energies are not problematic if there is no interaction term, such as $\frac{1}{4}\Lambda u^4$ in Eq. (2.46), or $\frac{\lambda}{4}(x^2 - y^2)^2$ in Eqs. (2.14), (2.22). Then the system behaves like a system of free oscillators, and it does not matter whether some of them have positive and some negative energies. Such a free system can be described by a Lagrangian in which all degrees of freedom have positive energies. Several authors [14–24] have found that the (free) Pais-Uhlenbeck oscillator can be described in terms of positive energies only. This is no longer so in the presence of an interaction. The Lagrangian (2.46) cannot be transformed into a Lagrangian for two mutually interacting positive energy oscillators [12, 28]. One of them necessarily has negative energy. It has been generally believed that such a system is unstable. But numerical calculations show that it is not necessarily so [25–28]. The system is stable for certain range of the initial velocity and the coupling constant Λ. But in Ref. [29] an unconditionally stable interacting system was found.

If instead of the Lagrangian (2.46) with the unbounded interaction potential $\frac{1}{4}\Lambda u^4$ we take a bounded potential, such as $\frac{1}{4}\Lambda \sin^4 u$, or $\frac{1}{4}\Lambda e^{-(x+y)^2}$, then numerical calculation show [28] that such system is stable in the variable u, whilst in the variables x, y is stable in the sense that their time derivatives (velocities) cannot escape into infinity, but approach finite values, whereas the x and y asymptotically behave as coordinates of a free particle and increase linearly with time.

2.3   Collision or scattering of two particles

Let us now involve into the game another particle and consider the collision or scattering of two particles. Confining to the case of two dimensions for each particle, let the Lagrangian be

$$Case\ I \quad L = \frac{1}{2}\left(\dot{x}_1^2 - \dot{y}_1^2 + \dot{x}_2^2 - \dot{y}_2^2\right) - V(x_1 - x_2, y_1 - y_2), \quad (2.49)$$

from which we obtain the following equations of motion

$$\ddot{x}_1 + \frac{\partial V}{\partial x_1} = 0\ , \quad \ddot{y}_1 - \frac{\partial V}{\partial y_1} = 0\ , \quad (2.50)$$

$$\ddot{x}_2 + \frac{\partial V}{\partial x_2} = 0\ , \quad \ddot{y}_2 - \frac{\partial V}{\partial y_2} = 0\ . \quad (2.51)$$

This system of differential equations can also be numerically integrated by using Mathematica. In Fig. 2.13 we show the results for

$$V = e^{-a[(x_1-x_2)^2+(y_1-y_2)^2]}\left[\frac{1}{2}[(x_1 - x_2)^2 - (y_1 - y_2)^2]\right.$$

$$\left. + \frac{k}{4}[(x_1 - x_2)^2 - (y_1 - y_2)^2]\right]. \quad (2.52)$$

We see that in this scenario the incoming velocities of the particles 1 and 2 can be small, whilst the velocities of the outgoing particles can be high. For instance, the component $\dot{x}_1^2/2$ of the kinetic energy increases after the collision by a factor of about 50. We have tried many other scenarios

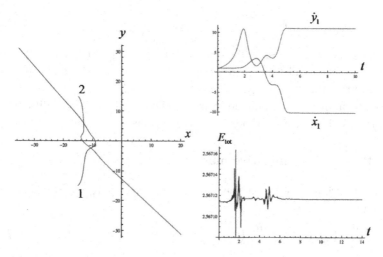

Fig. 2.13 Collision of particle 1 and particle 2 for Case I in the presence of the potential (2.52) for $a = 0.01$ and $k = 0.1$. The initial data are $\dot{x}_1(0) = 1.09$, $\dot{y}_1 = 0.97$, $x_1 = -15$, $y_1 = -15$, $\dot{x}_2 2(0) = 1$, $\dot{y}_1 2 = -1$, $x_2 = -15$, $y_2 = 15$. In the left plot are shown the particles' trajectories, whilst in the right upper plot is the velocity of particle 1, and in the lower plot the total energy as function of time.

with different initial conditions and potentials, and in all cases the colliding particles behaved normally, apart from the eventual increase of the velocities from small to high absolute values. Calculations also confirm validity of the conservation of energy and momentum:

$$\frac{1}{2}\left(\dot{x}_1^2 - \dot{y}_1^2 + \dot{x}_2^2 - \dot{y}_2^2\right) + V(x_1 - x_2, y_1 - y_2) = C, \qquad (2.53)$$

$$\dot{x}_1 + \dot{x}_2 = C_x, \qquad (2.54)$$

$$\dot{y}_1 + \dot{y}_2 = C_y. \qquad (2.55)$$

An example for energy conservation is shown in Fig. 2.13 (left, down).

In the case of the Lagrangian (2.49) particle 1 has one degree of freedom, x_1, with positive kinetic energy, and one degree of freedom, y_1, with negative kinetic energy. The same holds for particle 2.

Another possibility is the Lagrangian

Case II $\qquad L = \frac{1}{2}\left(\dot{x}_1^2 + \dot{y}_1^2 - \dot{x}_2^2 - \dot{y}_2^2\right) - V(x_1 - x_2, y_1 - y_2), \qquad (2.56)$

which gives the equations of motion

$$\ddot{x}_1 + \frac{\partial V}{\partial x_1} = 0 \,, \qquad \ddot{y}_1 + \frac{\partial V}{\partial y_1} = 0 \,, \qquad (2.57)$$

$$\ddot{x}_2 - \frac{\partial V}{\partial x_2} = 0 \;, \qquad \ddot{y}_2 - \frac{\partial V}{\partial y_2} = 0 \;. \tag{2.58}$$

In this case, both degrees of freedom, x_1 and y_1, of particle 1 have positive kinetic energies, whilst the degrees of freedom x_2 and y_2 of particle 2 have both negative kinetic energy.

Behavior of such system for various choices of initial conditions and potentials V is shown in in next two subsections.

2.4 Positive and negative masses

The Lagrangian (2.56) is a special case, for $m_1 = 1$, $m_2 = -1$, of the Lagrangian

$$\frac{m_1}{2}\left(\dot{x}_1^2 + \dot{y}_1^2\right) + \frac{m_2}{2}\left(\dot{x}_2^2 + \dot{y}_2^2\right) - V(x_1 - x_2, y_1 - y_2). \tag{2.59}$$

We are thus considering a model in which the mass of a particle can be either positive or negative. Such models have been much investigated in the literature, either within the context of gravity [30] or other dynamical system (see. e.g., Ref. [31–34]).

In the case of gravity, the Lagrangian (2.59), of course, has to be extended to three spatial dimensions, and the potential is $V = -G\frac{m_1 m_2}{r_{12}}$, where $r_{12} = \sqrt{(x_1 - x_2)^2 + (y_1 - y_2)^2 + (z_1 - z_2)^2}$. in our calculations we suppressed the third dimension and considered the Lagrangian (2.59) with

$$V(x_1 - x_2, y_1 - y_2) = -\frac{m_1 m_2}{\sqrt{(x_1 - x_2)^2 + (y_1 - y_2)^2}}. \tag{2.60}$$

After checking that our procedure works well for positive masse m_1, m_2, we then performed calculations for various positive and negative masses. Our results are in agreement with the results provided by Choi in the papers[2] [35–37] and confirm the known result that equal masses of opposite signs, starting with zero velocity, both move into the same direction with uniform acceleration, which goes on for ever (Fig. 2.14 up). Such behavior holds if we take exactly equal absolute values of masses, and exactly zero relative velocity, which is an unrealistic situation. If the relative velocity is different from zero, then in the case of *exactly* head on collision, which also is physically unrealistic situation, the acceleration is not uniform, but increases towards infinity (Fig. 2.14 down). In a realistic case, when a collision is not exactly head on, then however small is the deviation, the velocity of both particles reaches a finite value (Fig. 2.15).

---

[2]See also the simulations at https://www.youtube.com/watch?v=MZtS7cBMIc4.

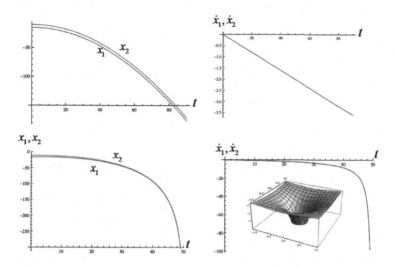

Fig. 2.14   Exactly head on collision of two particles in the presence of the gravitational interaction derived from the potential (2.60) for $m_1 = 1$ and $m_2 = -1$. Up: $x_1'(0) = 0, y_1'(0) = x_2'(0) = y_2'(0) = 0$. Down: $x_1'(0) = 0.1, y_1'(0) = x_2'(0) = y_2'(0) = 0$. In both cases the initial positions are the same: $x_1(0) = -15, y_1(0) = 0, x_2(0) = -10, y_2(0) = 0$.

If the absolute masses are not exactly equal, but slightly different, then the behavior of the system depends on which one is smaller. If the positive mass m_1 is smaller than $|m_2|$, then there is no runaway behavior of velocity even in the case of exactly head on collision or exactly zero relative velocity. If $m_1 > |m_2|$, the velocity escapes into infinity similarly as in Fig. 2.14 down.

If a collision is not head on, then a typical trajectory for $m_1 = 1$, $m_2 = -1$ is shown in Fig. 2.16, left up. For comparison, in Fig. 2.16, left down, is shown for the same initial data the case of equal positive masses, $m_1 = m_2 = 1$, which gives the elliptical trajectories.

Next, if we take equal absolute masses, $m_2 = -m_1$, and modify the potential so that it becomes bounded, e.g.,

$$V = -\frac{m_1 m_2 e^{-\frac{a}{r^2}}}{r}, \qquad (2.61)$$

where $r \equiv r_{12} = \sqrt{(x_1 - x_2)^2 + (y_1 - y_2)^2}$, then the runway motion, known also under the name "diametric drive", occurs only if the initial velocities of both particles are zero (i.e., the relative velocity is zero). But if, e.g., particle 1 has non vanishing initial velocity (however small), then the runaway behavior in a head on collision does not last for ever. At a certain moment there is the turning point when the (absolute) increase of velocity

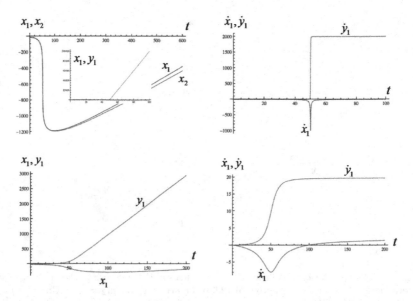

Fig. 2.15 A slight deviation from a head on collision brings no runaway behavior of velocity. Up: $x_1(0) = -15$, $y_1(0) = 0.01$, $x_2(0) = -10$, $y_2(0) = 0$. Down: $x_1(0) = -15$, $y_1(0) = 1$, $x_2(0) = -10$, $y_2(0) = 0$ In both cases the initial velocities are the same: $x_1'(0) = 0.1, y_1'(0) = x_2'(0) = y_2'(0) = 0$, and $m_1 = 1$ and $m_2 = -1$.

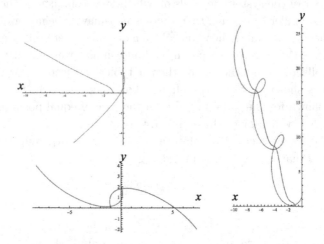

Fig. 2.16 Trajectories in the (x, y)-space for the initial data $x_1(0) = -1$, $y_1 = 0$, $x_2 = 0$, $y_2 = 2$, $\dot{x}_1 = -0.1$, $\dot{y}_1 = 0.6$, $\dot{x}_2 = 0.2$, $\dot{y}_2 = -0.6$, for three different choices of masses: $m_1 = 1$, $m_2 = -1$ (left up), $m_1 = m_2 = 1$ (left down), $m_1 = 5$, $m_2 = -1$ (right).

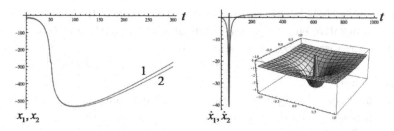

Fig. 2.17 Exactly head on collision in the presence of a damped gravitational potential (2.61) with $a = 0.01$ for $m_1 = 1$, $m_2 = -1$, and the initial data $x_1(0) = -15$, $y_1 = 0$, $x_2 = -10$, $y_2 = 0$, $\dot{x}_1 = 0.1$, $\dot{y}_1 = 0$, $\dot{x}_2 = 0$, $\dot{y}_2 = 0$, which are the same as in Fig. 2.14 down.

reverses, and the velocity starts decreasing and finally reaches a constant value (Fig. 2.17). Again the bounded (limited, not going to infinity) potential stabilizes the system. Runaway behavior takes place only in the case when the initial relative velocity between the particles is exactly zero. But such a situation is unrealistic, because in attempting to achieve such initial condition, the particles 1 and 2 must be transported by non-gravitational means close to each other and then manage to set their relative velocity to zero, but realistically this cannot be exactly zero. Thus no absurd runaway behavior occurs if we consider a realistic situation in which the potential is bounded and the relative velocity between the particles is not taken exactly zero.

2.5 Discussion

Our studies of the systems with negative energies have shown that negative energies are not problematic at al. It is usually uncritically taken for granted that such systems are unstable, without specifying what exactly is meant by their instability and under which conditions it occurs. Namely, also a system with positive kinetic energy is unstable at the top of an upside down potential. If such a potential has no bottom, then the system rolls down, and its kinetic energy increases towards infinity. We know that this is an unrealistic situation because a realistic potential has a bottom; therefore, we do not consider a system with positive kinetic energies problematic. But this situation is just the opposite of an analogous situation that involves a system with negative kinetic energy and a potential bounded from below, increasing into infinity. Such a system also exhibits unstable behavior; it rolls upward the potential while its kinetic energy approaches minus infinity. As a positive energy system is unstable at the top of a potential, so a negative energy system is unstable at the bottom of a potential. Because a realistic potential is bounded from below and from above, a system in the presence of such a potential is stable in the sense that its velocity remains finite.

We have seen that systems with negative kinetic energies allow for various scenarios in which the initial velocities are low, while the finite velocities are high. Such is a possible outcome of a collision of two particles, or scattering of a particle on a fixed potential. A question arises as to whether such high energy outgoing particles could be utilized to produce a useful work, either directly, or indirectly by heating a bulk of the surrounding material. The arrangement is very appealing: we start with a system of low energy particles, and end with a system of high energy particles.We must stress that such arrangement has nothing to do with a "perpetuum mobile", neither of the first, nor of the second kind. Namely, it is not a perpetuum mobile of the first kind, because during all the process, energy and momentum are strictly conserved. The increased absolute kinetic energy of the outgoing particles comes from the relative potential energy of the colliding particles. It is also not a perpetuum mobile of the second kind, because no "canalization" or "stirring" of the thermal energy into useful work is employed. But a question remains whether such arrangement is in agreement with the second law of thermodynamics. Namely, it could be that in the thermodynamic equilibrium the incoming flux of low energy particles is disturbed by the outgoing particles to such extent that no net flux of high energy particles exits from the arrangement. But there could also exist some ingenious ways to circumvent the above problem, so that the beams of incoming particles would not be disturbed by the beams of outgoing particles. In any case, as show our calculations, the outcomes are very sensitive to initial conditions: if they are slightly different, the energies of the outgoing particles are not significantly higher from those of the incoming particles. Therefore experimental reproductions of such effects would require precise control of many parameters.

An example of a higher derivative system is the Pais-Uhlenbeck oscillator. It is a toy model for a higher derivative gravity, and for any other system that can be described by a higher derivative theory. It is feasible to envisage that there exist or can be fabricated materials that give rise to the phenomena, e.g., acoustic waves, whose description can be done in terms of fields satisfying a higher derivative action principle. Today, many kinds of so called *metamaterials* are known with unusual physical properties, including negative index of refraction, negative effective mass density[3], or exhibiting non linear response to an external influence. This is a fast growing field of research with numerous possible applications, and my suggestion here is to explore, amongst others, also the possibility of constructing a metematerial behaving in accordance with a variant of the action principle (2.48) in which the potential term is bounded from below and above.

---

[3] A diametric drive acceleration due to the interaction between positive and negative (effective) mass for pulses propagating in a nonlinear mesh lattice was obsevred in the experiments performed by Wimmer et al. [38].

Chapter 3

Upon Quantization — Ghosts or Negatives Energies?

In the previous chapter we reviewed various approaches to the problem of negative energies that occur in higher derivative theories, a toy model for them being the Pais-Uhlenbeck (PU) oscillator. We have noted the well-known fact that negative energies themselves are not problematic, if there are no interactions between positive and negative energy degrees of freedom. Problems are expected to arise in the presence of interactions. Therefore, the approaches in which the authors show how the description of the PU oscillator in terms of an indefinite Hamiltonian can be replaced with a positive definite Hamiltonian, do not solve the problem [15–19]. That is, such a reformulation of a free PU oscillator does not work for an interacting PU oscillator. For instance, if one attempts to reformulate a self-interacting PU oscillator in terms of a positive definite Hamiltonian, the Lagrangian acquires additional nonlinear terms that are not present in the original Lagrangian [12]. Therefore, an interacting PU oscillator, as a higher derivative system, has to be described in terms of the Ostrogradski or an equivalent formalism that contains negative energies. In Refs. [12,25–28] it was shown that if the potential is *unbounded* either from below or from above, then such systems are not necessarily unstable; they can be either stable or unstable, depending on the values of the coupling constant and the initial velocity. However, upon quantization such system would become unstable because of the tunnelling effect.

But unbounded potentials are not physically realistic. For instance, a harmonic oscillator is an idealization, not realized in nature. An actual oscillator always has a potential that is bounded. In the case of systems whose energies are positive, the potential is required to be bounded from below, but usually it is taken for granted that it need not be bounded from above, despite that a physically realistic system is bounded from above as well. When considering higher derivative theories that contain negative energies, if we assume the presence of an unbounded potential, then the systems described by such theories are not stable in general. As shown in Refs. [12, 25–28], they can be stable within certain ranges of initial conditions and coupling constant, but they cannot be absolutely stable, i.e., stable for all values of those parameters. However, as we have pointed out, a physically realistic potential has to be bounded not only from bellow, but also from above. Then even the systems whose energies can be positive or negative, are absolutely stable [28, 29]. When such systems are quantized, they remain stable.[1]

So far we have considered higher derivative theories at the classical level. Now we are going to investigate what happens if we quantize the classical theories that involve negative energies. Common wisdom is that such theories contain ghosts, i.e., the states with negative norms. However, it was pointed out that a correct quantization, satisfying the correspondence principle [43–46], should not involve ghosts, but states with negative energies. We are going now to review[2], first on simpler and then on more involve examples, how to quantize systems whose degrees of freedom reside in spaces with indefinite metric and hence contain both positive and negative energies.

3.1 Illustrative example: The system of two equal frequency oscillators

Let us consider a toy model, described by the following Lagrangian and the Hamiltonian[3]

$$L = \frac{1}{2}(\dot{x}^2 - \dot{y}^2) - \frac{1}{2}\omega^2(x^2 - y^2), \qquad (3.1)$$

$$H = p_x\dot{x} + p_y\dot{y} - L = \frac{1}{2}(p_x^2 - p_y^2) + \frac{\omega^2}{2}(x^2 - y^2). \qquad (3.2)$$

The Hamilton equations of motion are

$$\dot{x} = \{x, H\} = \frac{\partial H}{\partial p_x} = p_x \, , \qquad \dot{y} = \{y, H\} = \frac{\partial H}{\partial p_y} = -p_y \qquad (3.3)$$

---

[1]An alternative approach to the interacting higher derivative systems, including the PU oscillator has, been considered in Refs. [39–42].

[2]The next two sections are taken and modified from Ref. [12]

[3]Though this system, because of the degeneracy $\omega_1^2 = \omega_2^2 = \omega^2$ is not equivalent to the PU oscillator, it serves well for illustration of the main points. It can be straightforwardly generalized to the case of unequal frequencies $\omega_1^2 \neq \omega_2^2$ of the two oscillators.

$$\dot{p}_x = \{p_x, H\} = -\frac{\partial H}{\partial x} = -\omega^2 x \ , \qquad \dot{p}_y = \{p_y, H\} = -\frac{\partial H}{\partial y} = \omega^2 y \quad (3.4)$$

where the Poisson brackets are defined as usual,

$$\{x, p_x\} = 1 \ , \qquad \{y, p_y\} = 1. \quad (3.5)$$

In the quantized theory we have commutators

$$[x, p_x] = i \ , \qquad [y, p_y] = i. \quad (3.6)$$

Introducing

$$c_x = \frac{1}{\sqrt{2}}(\sqrt{\omega}\, x + \frac{i}{\sqrt{\omega}}\, p_x) \ , \quad c_x^\dagger = \frac{1}{\sqrt{2}}(\sqrt{\omega}\, x - \frac{i}{\sqrt{\omega}}\, p_x) \quad (3.7)$$

$$c_y = \frac{1}{\sqrt{2}}(\sqrt{\omega}\, y + \frac{i}{\sqrt{\omega}}\, p_y) \ , \quad c_y^\dagger = \frac{1}{\sqrt{2}}(\sqrt{\omega}\, y - \frac{i}{\sqrt{\omega}}\, p_y) \quad (3.8)$$

we have

$$[c_x, c_x^\dagger] = 1 \ , \qquad [c_y, c_y^\dagger] = 1, \quad (3.9)$$

$$[c_x, c_y] = 0 \ , \qquad [c_x^\dagger, c_y^\dagger] = 0. \quad (3.10)$$

The Hamiltonian (3.2) can be written as

$$H = \omega\, (c_x^\dagger c_x + c_x c_x^\dagger - c_y^\dagger c_y - c_y c_y^\dagger). \quad (3.11)$$

Defining the vacuum according to

$$c_x|0\rangle = 0 \ , \qquad c_y|0\rangle = 0 \quad (3.12)$$

the Hamiltonian becomes

$$H = \omega\, (c_x^\dagger c_x - c_y^\dagger c_y). \quad (3.13)$$

All states have positive norms, e.g.,

$$\langle 0|cc^\dagger|0\rangle = \langle 0|[c, c^\dagger]|0\rangle = \langle 0|0\rangle = 1, \quad (3.14)$$

regardless of whether the subscript of c, $c^\dagger$ is x or y, because both commutators in Eq. (3.9) are equal to 1.

In the coordinate representation the momenta are

$$p_x = -i\frac{\partial}{\partial x} \ , \qquad p_y = -i\frac{\partial}{\partial y}, \quad (3.15)$$

whereas the vacuum state and its defining equations become

$$\langle x, y|0\rangle = \psi_0(x, y), \quad (3.16)$$

$$\frac{1}{2}\left(\sqrt{\omega}\,x + \frac{1}{\sqrt{\omega}}\,\frac{\partial}{\partial x}\right)\psi_0(x,y) = 0, \qquad (3.17)$$

$$\frac{1}{2}\left(\sqrt{\omega}\,y + \frac{1}{\sqrt{\omega}}\,\frac{\partial}{\partial y}\right)\psi_0(x,y) = 0. \qquad (3.18)$$

The solution is

$$\psi_0 = \frac{2\pi}{\omega}e^{-\frac{1}{2}\omega(x^2+y^2)}, \qquad (3.19)$$

and it has *positive norm*

$$\int \psi_0^2 \, dx \, dy = 1. \qquad (3.20)$$

The vacuum $\psi_o(x,y)$ and the states $\psi(x,y)$ excited by successive action of the operators $c_x^\dagger, c_y^\dagger$ on $\psi_0(x,y)$ are not invariant under the hyperbolic rotations in the (x,y)-space, under which the Lagrangian (3.1) and the Hamiltonian (3.2) are invariant. However, even if solutions are not invariant, the theory is covariant, because the set of possible solutions in a new reference frame corresponds to the set of possible solutions in the old reference frame. Thus, though in a new frame S' the vacuum (3.19) becomes

$$\psi_0(x',y') = \frac{2\pi}{\omega}e^{-\frac{1}{2}\omega\left((x'\cosh\alpha - y'\sinh\alpha)^2 + (-x'\sinh\alpha + y'\cosh\alpha)^2\right)}, \qquad (3.21)$$

there also exists the solution

$$\psi_0'(x',y') = \frac{2\pi}{\omega}e^{-\frac{1}{2}\omega(x'^2+y'^2)}, \qquad (3.22)$$

which has the same form as $\psi_0(x,y)$ of Eq. (3.19). The same is true for all excited states.

Such a model can be straightforwardly generalized to higher dimensional spaces with signature (r,s). The Lagrangian and the Hamiltonian are

$$L = \frac{1}{2}\dot{x}^a\dot{x}_a - \frac{1}{2}\omega^2 x^a x_a, \qquad (3.23)$$

$$H = \frac{1}{2}p^a p_a + \frac{1}{2}\omega^2 x^a x_a, \qquad (3.24)$$

where indices are lowered and raised with the metric η_{ab} and its inverse η^{ab}. The momentum is $p_a = \partial L/\partial \dot{x}^a = \dot{x}_a = \eta_{ab}\dot{x}^b$. Upon quantization we have

$$[x^a, p_b] = i\delta_b^a. \qquad (3.25)$$

The creation/annihilation operators defined according to

$$c^a = \frac{1}{\sqrt{2}} \left(\sqrt{\omega} x^a + \frac{i}{\sqrt{\omega}} p_a \right), \tag{3.26}$$

$$c^{a\dagger} = \frac{1}{\sqrt{2}} \left(\sqrt{\omega} x^a - \frac{i}{\sqrt{\omega}} p_a \right), \tag{3.27}$$

are a generalization of the operators $c_x, c_x^\dagger, c_y, c_y^\dagger$ given in Eqs. (3.7), (3.8). They satisfy the commutation relations

$$[c^a, c^{b\dagger}] = \delta^{ab}, \quad [c^a, c^b] = [c^{a\dagger}, c^{b\dagger}] = 0. \tag{3.28}$$

The Hamiltonian is

$$H = \frac{1}{2} \omega \left(c_a^\dagger c^a + c^a c_a^\dagger \right) = \omega \left(c_a^\dagger c^a + \frac{r}{2} - \frac{s}{2} \right). \tag{3.29}$$

If vacuum is defined as

$$c^a |0\rangle = 0, \tag{3.30}$$

then the vacuum expectation value of H is

$$\langle H \rangle = \frac{\omega}{2} (r - s), \tag{3.31}$$

which vanishes if $r = s$, that is when the signature is neutral.

In the literature an alternative definition of annihilation/creation operators is usually employed, namely

$$a^a = \frac{1}{2} \left(\sqrt{\omega} x^a + \frac{i}{\sqrt{\omega}} p^a \right), \tag{3.32}$$

$$a^{a\dagger} = \frac{1}{2} \left(\sqrt{\omega} x^a - \frac{i}{\sqrt{\omega}} p^a \right), \tag{3.33}$$

that satisfy

$$[a^a, a_b^\dagger] = \delta^a{}_b, \quad [a^a, a^{b\dagger}] = \eta^{ab}. \tag{3.34}$$

There are two possible definitions of vacuum:
Definition I. This is the usual definition,

$$a^a |0\rangle = 0. \tag{3.35}$$

The Hamiltonian

$$H = \frac{1}{2} \omega \left(a^{a\dagger} a_a + a_a a^{a\dagger} \right) = \omega \left(a^{a\dagger} a_a + \frac{r}{2} + \frac{s}{2} \right), \tag{3.36}$$

acting on the states created by $a^{a\dagger}$ has always *positive eigenvalues*. But the states corresponding to negative signature have *negative norms*, and are therefore called *ghosts*.

Definition II. This is the Cangemi-Jackiw-Zwiebach definition [43, 44], discussed in Refs. [45, 46]. We split the operators into the positive and negative signature parts,

$$a^a = (a^{\bar{a}}, a_{\underline{a}}) \,, \qquad \bar{a} = 1, 2, ..., r \,;$$

$$\underline{a} = r + 1, r + 2, ..., r + s. \qquad (3.37)$$

Then the Hamiltonian operators (3.36) which can be written as

$$H = \omega \left(a^{\bar{a}\dagger} a_{\bar{a}} + a_{\underline{a}} a^{\underline{a}\dagger} + \frac{r}{2} - \frac{s}{2} \right), \qquad (3.38)$$

has positive and negative eigenvalues. There are no negative norm states. If the signature is neutral, $r = s$, then the vacuum energy vanishes.

Analogous happens if we extend the system (3.37) to infinite degrees of freedom, for instance, to a system of scalar fields, living in a field space with neutral metric [45, 47]. Then the vacuum energy is not infinite, but vanishes. Consequently, if such a system is coupled to gravity, the notorious cosmological constant problem does not arise; the cosmological constant is zero. The observed accelerated cosmological expansion, ascribed to dark energy, and possibly due to a small cosmological constant, is thus an effect whose explanation still remains to be found.

The quantization of the system (3.23) according to Definition II has the correct classical limit, and satisfies the correspondence principle. Woodard [46] argues that this is the correct quantization, whereas the quantization according to Definition I is incorrect.

Unfortunately, many authors who have been aware of the Definition II, considered such a system as problematic anyway, because in physically realistic situations there are interactions between positive and negative energy degrees of freedom. According to the prevailing opinion, such a system is necessarily unstable in the presence of interactions. But it has turned out that this is not true. Behaviour of the system (3.1) in the presence of various interactions has been studied in Ref. [11, 12, 48], where it was found that for suitable interactions the system is stable. This was reviewed and further elaborated in the previous chapter.

3.2   Interacting quantum oscillator

In the Schrödinger coordinate representation the quantum oscillator is described by the wave function $\psi(t, x, y)$, satisfying the Schrödinger equation

$$i \frac{\partial \psi}{\partial t} = H \psi, \qquad (3.39)$$

with

$$H = \frac{1}{2}\left(-\frac{\partial}{\partial x^2} + \frac{\partial}{\partial y^2}\right) + V(x,y). \tag{3.40}$$

Reproducing here the procedure considered in Ref. [11], let us expand the wave function in terms of the basis functions of the $2D$ harmonic pseudo-Euclidean oscillator, which are the same as those of the Euclidean oscillator:

$$\psi = \sum_{m,n=0}^{\infty} c_{mn}(t)\,\psi_{mn}. \tag{3.41}$$

Here $\psi_{mn} = \frac{1}{\pi^{1/4}\sqrt{2^n n!}}\frac{1}{\pi^{1/4}\sqrt{2^m m!}} H_n(x)H_m(y)e^{-(x^2+y^2)/2}$ are orthonormal eigenfunctions (with positive or negative energies), of the Hamiltonian

$$H_0 = \frac{1}{2}\left(-\frac{\partial}{\partial x^2} + \frac{\partial}{\partial y^2}\right) + \frac{1}{2}(x^2 - y^2), \tag{3.42}$$

$H_n(x)$ being the Hermit polynomials.

Using the matrix elements

$$H_{mn;rs} = \int dx\,dy\,\psi_{mn}^* H\psi_{rs}, \tag{3.43}$$

the Schrödinger equation (3.39) can be rewritten as

$$i\dot{c}_{mn} = \sum_{rs} H_{mn;rs}\,c_{rs}. \tag{3.44}$$

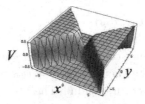

Fig. 3.1   The potential used in the calculations of the quantum pseudo-Euclidean harmonic oscillator described by Eqs. (3.39), (3.40).

We will investigate the case of the potential (see Fig. 3.1)

$$V(x,y) = \frac{1}{2}\varepsilon\left(1 - e^{-\varepsilon(x^2 - y^2)}\right), \tag{3.45}$$

where $\epsilon = \text{sign}\,(x^2 - y^2)$, i.e., $\epsilon = 1$, if $x^2 - y^2 > 0$, $\epsilon = -1$, if $x^2 - y^2 < 0$, and $\epsilon = 0$, if $x^2 - y^2 = 0$. Such a potential is a 2D model of a potential that could eventually occur in a three or higher dimensional world, where potentials running into infinity are unrealistic.

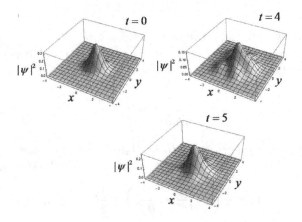

Fig. 3.2   The plot of $|\psi(t, x, y)|^2$, calculated for the initial conditions $c_{00}(0) = 1$, $c_{01}(0) = C_{10}(0) = c_{11}(0) = \ldots = 0$, at different values of the time t.

Instead of the system (3.44) of infinite number of first order differential equations, let us consider a finite system with $m, n = 0, 1, 2, \ldots, N$. Then we can numerically solve the system and calculate the coefficients $C_{mn}(t)$ for given initial conditions. So we obtain the wave functions

$$\psi(t, x, y) = \sum_{m,n=0}^{N} c_{mn}(t)\, \psi_{mn}, \tag{3.46}$$

and its absolute square, $|\psi(t, x, y)|^2$.

In Fig. 3.2 we show the result for $N = 4$, and the initial condition $c_{00} = 1$, with the remaining coefficients being equal to zero. We see that with increasing time t, the vacuum $\psi(0) = \psi_{00}$ gradually decays, and, after a while (at $t \approx 5$) occurs again. The wave function thus oscillates between the vacuum and a decayed vacuum.

In Fig. 3.3 we show the calculation of $|\psi(t, x, y)|^2$ for the initial conditions $c_{01}(0) = 1/\sqrt{2}$, $c_{10}(0) = 1/\sqrt{2}$, $c_{00}(0) = 0$, $c_{11}(0) = 0$, $c_{12}(0) = 0, \ldots, c_{44}(0) = 0$. Initially, we have two peaks, one for the positive energy excitation, $c_{10}(0)$, and the other one for the negative energy excitations, $c_{01}(0)$. The system then oscillates as shown in Fig. 3.3.

Such a wave function (3.46), of course, is not a solution of the Schrödinger equation (3.39) for H given in Eqs. (3.40), (3.45). Since $c_{mn}(t)$, $m, n = 0, 1, 2, 3, 4$, are solution of the truncated system (3.44),

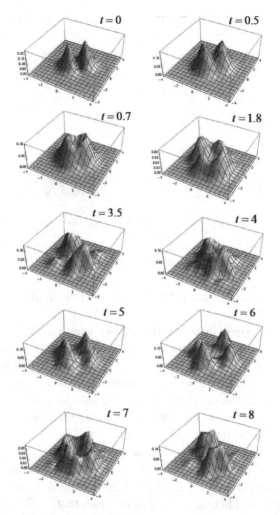

Fig. 3.3 The plot of $|\psi(t,x,y)|^2$, calculated for the initial conditions $c_{01}(0) = C_{10}(0) = 1/\sqrt{2}$, $c_{00}(0) = c_{11}(0) = c_{12}(0) = \ldots = 0$, at different values of the time t.

also $\psi(t,x,y)$ of Eq. (3.46) is a solution of a truncated system, presumably of the Schrödinger equation on the lattice. A realistic lattice is a discrete set of closely separated points corresponding to the system, not of the 5×5 equations (3.44), but of the $N \times N$ equations, with N very large. The number N is related to the maximum absolute energy of the oscillator excitations that, in turn, is related to the minimum distance (cutoff). Regardless of how large is N, and how small is the cutoff distance, the

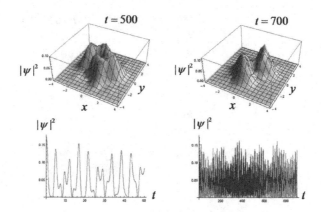

Fig. 3.4 Up: The plot of $|\psi(t,x,y)|^2$, calculated for the initial conditions $c_{01}(0) = C_{10}(0) = 1/\sqrt{2}$, $c_{00}(0) = c_{11}(0) = c_{12}(0) = \ldots = 0$, at two very large values the time, $t = 500$ and $t = 700$. Low: The plot of $|\psi(t, x = 0, y = 50)|^2$, as a function of t, calculated for the same initial conditions as above.

examples of Figs. 3.2–3.4 indicate that there are oscillations, and that the system is thus stable.

According to common belief, the Planck length is the cutoff distance under which it is no longer possible to probe spacetime distances. Our spacetime is then like a huge lattice. Then also the corresponding Clifford space, C, to be considered in next chapters, is like a lattice. The space $M_{1,1}$, considered in the previous section, is a subspace of C. Instead of the infinite system of differential equations (3.44), we then have a finite, though very big, system of differential equations.

If in our calculations we replace the potential (3.45) with the potential (2.28) considered in previous chapter, the results of calculations are similar to those shown in Figs. 3.2–3.4. This is so even if we set the damping constant a to zero. Then the potential becomes that of Eq. (2.22), i.e., unbounded, and yet nothing very drastically happens with the wave function. Instead of Figs. 3.2 and 3.4 we obtain Fig. 3.5 and 3.6. For control, we also performed calculation for the coupled Euclidean oscillators whose action has the same sign in front of the x and the y terms, so that there are no negative energy states. As expected, the vacuum state remains unperturbed, and the probability density does not oscillate between a Gaussian shape and a shape with extra bumps like in the middle of Fig. 3.5. Analogous hold also for the excited state starting with $c_{01} = c_{10} = 1/\sqrt{2}$.

Our calculations thus show that the quantum system with the Hamiltonian (3.40), (3.42) exhibit a decent behavior, in spite of the fact that the

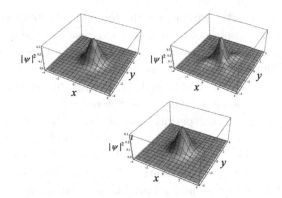

Fig. 3.5 The plot of $|\psi(t,x,y)|^2$, calculated for the initial conditions $c_{00}(0) = 1$, $c_{01}(0) = C_{10}(0) = c_{11}(0) = ... = 0$, at times $t = 0$, $t = 4$, and $t = 5$ (as in Fig. 3.2). The potential is now that of Eq. (2.28).

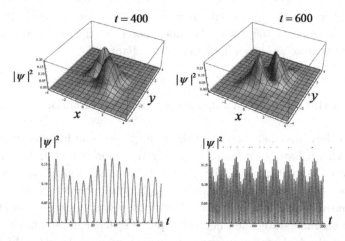

Fig. 3.6 Up: The plot of $|\psi(t,x,y)|^2$, calculated for the initial conditions $c_{01}(0) = C_{10}(0) = 1/\sqrt{2}$, $c_{00}(0) = c_{11}(0) = c_{12}(0) = ... = 0$, at two very large values the time, $t = 400$ and $t = 600$. Low: The plot of $|\psi(t, x = 0, y = 50)|^2$, as a function of t, calculated for the same initial conditions as above. The potential is now that of Eq. (2.28).

initial vacuum state "decays" in the sense that higher excited states take place. Nothing unusual happens with the probability density. In our truncated system in which we considered only the excited states from $n = 0$ to $n = 3$, $|\psi(t,x,y)|^2$ oscillated between a shape corresponding to a pure state and a state with a mixture of excited states. In Figs. 3.2–3.7 this manifested itself as occurrence of bumpy shapes. To see how the number of excited states considered influences the result, I repeated the calculation by diminishing the number of terms in the expansion of the wave function, and considered only the terms up to $n = 2$. For the vacuum initial state there was no difference; the plots remains the same, with the same period of oscillations between the bumps and no bumps. But for the initial state $c_{01} = c_{1,0} = 1/\sqrt{2}$, the plots reveal some difference.

A next step is therefore to perform the calculation by taking into account many excited states and thus approach to what would happen in the limit of infinite terms in the expansion (3.41) of the wave functions. So we will see whether the initial vacuum state still return with the same period, after passing bumpy states, into its original non bumpy state, or perhaps the period will become longer with increasing number of terms taken into account, and hence infinite for $n = \infty$.

Investigation of coherent states should confirm that the expectation values of wave packet behave as classical trajectories. This means that in case of unbounded potentials they will exhibit ever increasing expected velocities, whilst in the case of bounded potential the expected velocities will remain finite as shown in the previous chapter.

3.3   On the stability of higher derivative field theories

A higher derivative field theory of the form (2.48) is an infinite set of Pais-Uhlenbeck oscillators. If instead of the interaction potential $\lambda\phi^4$ we take a potential $V_1(\phi)$ that is bounded from below and form above, then just as in the case of the PU oscillator also such a field theoretic system is stable. However, now that we have infinite dimensional system, the number of available final states, and hence the phase space, can be infinite. In the studies of physical processes, one usually sums over final states. With both positive and negative energies, the phase space can be infinite. The formulas for transition and decay rates thus besides the transition amplitude squared contain the integration over the phase space. A common conclusion is that because in the case of higher derivative theories the phase space is infinite, vacuum would instantaneously decay into positive and negative energy particles, which means that such theories are not physically viable.

Such a conclusion has been questioned in Ref. [11]. When studying a physical system, for instance an excited nucleus, we usually do not measure the momenta of the decay particles, therefore in the transition probability formula

$$\text{Transition probability} = \int |S_{fi}|^2 d^3 \boldsymbol{p}_1 d^3 \boldsymbol{p}_2... \tag{3.47}$$

we integrate over them. If we measure the final state particle's momenta, the integration goes over a narrower portion of the phase space, which reduces the transition probability.

As an example let us consider the scalar fields described by the action

$$I = \frac{1}{2} \int d^4x [g^{\mu\nu} \partial_\mu \varphi^a \partial_\nu \varphi^b \gamma_{ab} + V(\varphi)]. \tag{3.48}$$

If the metric γ_{ab} in the space of the fields φ^a is indefinite with signature (r, s), then the latter action can be obtained as a generalization of the action (2.37) that comes from the higher derivative action (2.47) that is a model for the field action (2.48). Let the potential be of the form

$$V(\varphi) = m^2 \varphi^a \varphi^b \gamma_{ab} + \varphi^a \varphi^b \varphi^c \varphi^d \lambda_{abcd}, \tag{3.49}$$

or similar.

After performing the standard quantization procedure, the system is described by a state vector $|\Psi\rangle$ expanded in terms of the Fock space basis vectors, $|P\rangle \equiv |p_1 p_2 ... p_n\rangle$ that are eigenvectors of the free field Hamiltonian H_0 (without the interaction quartic term $\varphi^a \varphi^b \varphi^c \varphi^d \lambda_{abcd}$):

$$|\Psi\rangle = \sum |P\rangle \langle P|\Psi\rangle. \tag{3.50}$$

The system evolves according to the Schrödinger equation

$$i \frac{\partial |\Psi\rangle}{\partial t} = H|\Psi\rangle, \tag{3.51}$$

with the formal solution

$$|\Psi(t)\rangle = e^{-iH(t-t_0)} |\Psi(t_0)\rangle. \tag{3.52}$$

Because the signature of the field space metric γ_{ab} is (r, s), a Fock space state vector contains particles with positive and negative energies. A consequence is that if the initial state is the vacuum $|\psi(t_0)\rangle = |0\rangle$, it can evolve into a superposition of many particle states, because the transitions

$$\langle P|e^{-iH(t-t_0)}|0\rangle = \langle P|\Psi(t)\rangle \tag{3.53}$$

can satisfy the energy and momentum conservation.

The vacuum thus decays into a superposition of many particle states:

$$|\Psi(t)\rangle = \sum_{n=0}^{\infty} |p_1 p_2 ... p_n\rangle \langle p_1 p_2 ... p_n|\Psi(t)\rangle, \tag{3.54}$$

where $\langle p_1 p_2 ... p_n | \Psi(t) \rangle$ is the probability amplitude of observing at time t the many particle state $|p_1 p_2 ... p_n\rangle$. The probabilities that the vacuum decays into any of the states $|p_1\rangle$, $|p_1 p_2\rangle$, $|p_1 p_2 ... p_n\rangle$, ..., are not very different from each other, and, after a proper normalization, they sum to $1 - |\langle 0 | \Psi \rangle|^2$. We thus have

$$|\langle 0 | \Psi \rangle|^2 + \sum_{p_1} |\langle p_1 | \Psi \rangle|^2 + \sum_{p_1, p_2} |\langle p_1, p_2 | \Psi \rangle|^2$$

$$+ \sum_{p_1, p_2, ..., p_n} |\langle p_1, p_2, ..., p_n | \Psi \rangle|^2 + ... = 1. \quad (3.55)$$

This is so, because at any time t the system must be in one of the states $|0\rangle$, $|p_1\rangle$, $|p_1, p_2\rangle$, ..., $|p_1, p_2, ..., p_n\rangle$. Therefore the total probability of finding the system in any of those states is 1. However, the probability that vacuum decays into 2,4,6,8, or any finite number of particles is infinitely small in comparison with the probability that it decays into infinite numbers of particles, because such configurations occupy the vast majority of the phase space. Therefore, for an *outside observer*, who does not measure momenta of the particles, the vacuum $|0\rangle$ instantly decays into infinitely many particles. The usual reasoning in the literature then goes along the lines that because of such vacuum instability the theories involving ultrahyperbolic spaces, and higher derivative theories in particular, are not physically viable.

However, every reasoning is based on certain assumptions, often implicit or tacit. The reasoning against higher derivative field theories based on the instantaneous vacuum decay due to the infinite phase space tacitly assumes the existence of an observer, outside the system described by the higher derivative field theory, who does measure the number of particles and their momenta. But in reality, such an outside observer cannot exist. Within our universe there can be no observer to whom the higher derivative theory does not apply, if such a theory is the theory describing our universe. Every conceivable observer is thus a part of our universe. If so, (s)he is then coupled to the particles in the universe, and thus at least implicitly measures their number and momenta. For such an observer, there is no instantaneous vacuum decay for the reason discussed in Ref. [11], quoted below:

Let us consider a generalization of the field action (3.48). We can rewrite it in a more compact notation:

$$I = \frac{1}{2} \partial_\mu \varphi^{a(x)} \partial_\nu \varphi^{b(x')} \gamma^{\mu\nu}{}_{a(x)b(x')} - U[\varphi] \quad (3.56)$$

Here $\varphi^{a(x)} \equiv \varphi^a(x)$, where (x) is the continuous index, denoting components of an infinite dimensional vector. In addition, for every (x), the components are also denoted by a discrete index a. Altogether, vector components are denoted by the double index $a(x)$. Alternative notation, often used in the literature, is $\varphi^a(x) \equiv \varphi^{ax}$ or $\varphi^a(x) \equiv \varphi^{(ax)}$.

The action (3.56) may be obtained from a higher dimensional action

$$I_\phi = \frac{1}{2} \partial_\mu \phi^{A(x)} \partial_\nu \phi^{B(x')} \, G^{\mu\nu}{}_{A(x)B(x')}, \qquad (3.57)$$

where $A = (a, \bar{A})$, and $\phi^{A(x)} = (\phi^{a(x)}, \phi^{\bar{A}(x)})$. The higher dimensional metric is a functional of $\phi^{A(x)}$. Performing the Kaluza-Klein split,

$$G^{\mu\nu}{}_{AB} = \begin{pmatrix} \gamma^{\mu\nu}{}_{ab} + A_a{}^{\bar{A}} A_b{}^{\bar{B}} \bar{G}^{\mu\nu}_{\bar{A}\bar{B}}, & A_a{}^{\bar{B}} \bar{G}^{\mu\nu}_{\bar{A}\bar{B}} \\ A_b{}^{\bar{B}} \bar{G}^{\mu\nu}_{\bar{A}\bar{B}}, & \bar{G}^{\mu\nu}_{\bar{A}\bar{B}} \end{pmatrix},$$
$$(3.58)$$

where for simplicity we have omitted the index (x), we find,

$$I_\phi = \frac{1}{2} \partial_\mu \phi^{a(x)} \partial_\nu \phi^{b(x')} \gamma^{\mu\nu}{}_{a(x)b(x')}$$

$$+ \frac{1}{2} \partial_\mu \phi_{\bar{A}(x)} \partial_\nu \phi_{\bar{B}(x)} \bar{G}^{\mu\nu \bar{A}(x)\bar{B}(x)}. \qquad (3.59)$$

Identifying $\phi^{a(x)} \equiv \varphi^{a(x)}$, and denoting $\frac{1}{2} \partial_\mu \phi_{\bar{A}(x)} \partial_\nu \phi_{\bar{B}(x)} \bar{G}^{\mu\nu \bar{A}(x)\bar{B}(x)} = -U[\varphi]$, we obtain the action (3.56).

Thus, the field action(3.48) is embedded in a higher dimensional action with a metric $G^{\mu\nu}{}_{A(x)B(x)}$ in field space. A question arises as to which field space metric to choose. The lesson from general relativity tells us that the metric itself is dynamical. Let us therefore assume that this is so in the case of field theory [49] as well. Then (3.57) must be completed by a kinetic term, I_G, for the field space metric [49]. The total action is then

$$I[\phi, G] = I_\phi + I_G. \qquad (3.60)$$

According to such dynamical principle, not only the field ϕ, but also the metric of field space $G^{\mu\nu}{}_{A(x)B(x)}$ changes with the evolution of the system. This implies that also the potential $U[\varphi]$ of Eq. (3.56) changes with evolution, and so does the potential $V(\varphi)$, occurring in Eq. (3.48).

Let us assume that the action (3.60) describes the whole universe[4]. Then $|\psi(t)\rangle$ of Eq. (3.54) contains everything in such universe, including observers. There is no external observer, $\mathcal{O}_{\text{ext}}$, according to whom the coefficients $\langle p_1...p_n|\psi(t)\rangle$ in Eq. (3.54) could be related to the probability densities $|\langle p_1...p_n|\psi(t)\rangle|^2$ of finding the system in an n-particle states with momenta p_1, ..., p_n, $n = 0, 1, 2, 3, ..., \infty$. There are only inside observers, $\mathcal{O}$, incorporated within appropriate multiparticles states $|p_1...p_n\rangle$, $n = 0, 1, 2, 3, ..., \infty$, of the "universal" state $|\psi(t)\rangle$. According to the Everett interpretation of quantum mechanics [215–217], all states, $|0\rangle$, $|p_1p_2\rangle$,..., $|p_1...p_n\rangle$, ..., $n = 0, 1, 2, ..., \infty$, in the superposition (3.54) actually occur, each in a different world. The Everett interpretation is now getting increasing support among cosmologists (see, e.g., Ref. [50, 51]). For such an inside observer, $\mathcal{O}$, there is no instantaneous vacuum decay into infinitely many particles. For $\mathcal{O}$, at a given time t, there exists a configuration $|P\rangle$ of n-particles (that includes $\mathcal{O}$ himself), and a certain field potential $V(\varphi)$ (coming from $G^{\mu\nu}{}_{A(x)B(x)}$). At some later time, $t + \Delta t$, there exists a slightly different configuration $|P'\rangle$ and potential $V'(\varphi)$, etc. Because $\mathcal{O}$ nearly continuously measures the state $|\psi(t)\rangle$ of his universe, the evolution of the system is being "altered", due to the notorious "watchdog effect" or "Zeno effect" [52] of quantum mechanics (besides being altered by the evolution of the potential $V(\varphi)$). The peculiar behavior of a quantum system between two measurements has also been investigated in Refs. [53, 54].

In Discussion of the same reference [11] it is then written:

After having investigated how the theory works on the examples of the classical and quantum pseudo Euclidean

---

[4]Of course, such a model universe is not realistic, because our universe contains fermions and accompanying gauge fields as well.

oscillator, we considered quantum field theories. If the metric of field space is neutral, then there occur positive and negative energy states. An interaction causes transitions between those states. A vacuum therefore decays into a superposition of states with positive and negative energies. Because of the vast phase space of infinitely many particles, such a vacuum decay is instantaneous—for an external observer. But we, as a part of our universe, are not external observers. We are internal observers entangled with the "wave function" of the rest of our universe. According to the Everett interpretation of quantum mechanics, we find ourselves in one of the branches of the universal wave function. In the scenario with decaying vacuum, our branch can consist of a finite number of particles. Once being in such a branch, it is improbable that at the next moment we will find ourselves in a branch with infinitely many particles. For us, because of the "watchdog effect" of quantum mechanics, the evolution of the universal wave function relative to us is frozen to the extent that instantaneous vacuum decay is not possible.

Another important point is that the field potential in the action (3.48) need not be fixed during the evolution of the universe, but can change. Also, a realistic potential is not unbounded, it should be bounded from below and from above, which then presumably gives stability as we have shown that it is the case in the classical theory (see Sec. 4.2). Moreover, as pointed by Carroll et al. [55], decay rates of the processes with negative energy particles ("phantoms" in their terminology) would not diverge if there were a cutoff to the phase space integrals. Maximum momenta imply minimum length, and in Ref. [11] it was pointed out that such a finite system (though with many degrees of freedom) oscillates between the "vacuum" state and the very many (but not infinitely many) particle state.

A realistic description of our universe must include fermions as well. A well known fact is that fermions are just particular Clifford numbers (algebraic spinors or Clifford aggregates) [56–61]. An element of the Clifford algebra generated by the basis vectors of 4-dimensional spacetime can be represented by a 4 matrix, and spinors are just elements of one column (an "ideal" of a Clifford algebra). Because there are four columns, there are four kinds of spinors. In Ref. [11] we find the following discussion (after a lengthy formalism supporting it):

> Description of our universe requires fermions and accompanying gauge fields, including gravitation. According to the Clifford algebra generalized Dirac equation—Dirac-Kähler equation [62–64]—there are four sorts of the 4-component spinors, with energy signs a shown in Eq. (111) [of Ref. [11]]. The vacuum of such field has vanishing energy and evolves into a superposition of positive and negative energy fermions, so that the total energy is conserved. A possible scenario is that the branch of the superposition in which we find ourselves, has the sea of negative energy states of the first and the second, and the sea positive energy states of the third and forth minimal left ideal of $Cl(1,3)$. According to Ref. [60,67], the former states are associated with the familiar, weakly interacting particles, whereas the latter states are associated with mirror particles, coupled to mirror gauge fields, and thus invisible to us. According to the field theory based on the Dirac-Kähler equation, the unstable vacuum could be an explanation for Big Bang.

Although the concept of the Dirac vacuum as a sea of negative energy particles is nowadays considered as obsolete, it finds its natural place within the Clifford algebra description of spinors and spinor fields [11,60,67,68,91]. This will be discussed in more detail in the next three chapters. Here let us just mention that if a higher derivative field theory describes our universe, then as a possible scenario it predicts that an initial "bare" vacuum state evolved into the present state, which involves the sea of negative energy fermions of the 1st and 2nd ideal (column) of Cl(1,3), and the sea of positive energy fermions of the 3rd and 4th ideal, so that their total energies sum to zero. The initially "bare" vacuum thus decayed into infinitely many particles not completely, but only partially, because half of the available fermion states remained finite, namely the positive energy states of the 1st and 2nd (that form our visible universe, including us), and the negative energy states of the 3rd and 4th ideal (that according to Ref. [60,67,91]) form mirror particles. Our sector of particles has remained finite, because we live in it and we observe (or measure) it. In other words, we are composed of those particles and we observe the same kind of particles in our environment and measure (at least implicitly) their momenta. But in this scenario we have not measured the momenta of those other kinds of fermions which then formed the sea. So people when talking about an (instantaneous) vacuum decay were only "half right". Vacuum indeed (presumably instantaneously) decayed into infinitely many particles, but only into half of the available four kinds of (fermionic) particles.

3.4 Conclusion

On the example of a harmonic oscillator with one positive and one negative degree of freedom we have shown that quantization of such a system is straightforward in the absence of an interactive term. In order to satisfy the correspondence principle, the correct quantization does not involve ghost states, but negative states. In the presence of a coupling term between the positive and negative energies, calculations show that nothing special happens with the system if the potential is bounded from above and below.

The issue regarding the infinite decay rate of vacuum due to the infinite phase space in higher derivative fields theories has also been clarified. The integration over phase space has to be performed when momenta of decaying products are not measured. If they are measured, then the integration does not run over the entire infinite phase space, but only over a small finite part of it, and therefore the decay rates are not infinite. Assume that our universe is governed by a higher derivative field theory. Then we observers living in such a universe are part of one of the infinitely many components of such a decayed vacuum, and we in fact implicitly measured momenta of the particles in our universe, therefore for us there was no instantaneous vacuum decay. Moreover, if there is a momentum cutoff due to the minimal length, the phase space is not infinite anyway. We conclude that higher derivative theories are physically viable with many important physical implications for further development of quantum gravity and its unification with the rest of physics.

Chapter 4

Transformations of Spinors

According to the generally accepted view spinors are objects that transform under rotations differently than vectors. However, it is also known by not so small group of experts that spinors can be considered as particular elements of Clifford algebras. If so, they must have the same transformation properties as any other Clifford number. This implies that spinor and vectors can both be transformed by the same set of transformations. In general, a Clifford number Φ transforms into a new Clifford number Φ' according to $\Phi \rightarrow \Phi' = R\,\Phi\,S$, i.e., by the multiplication from the left and from the right by two Clifford numbers R and S. We study the case of $Cl(1,3)$, which is the Clifford algebra of the Minkowski spacetime. Depending on choice of R and S, there are various possibilities, including the transformations of vectors into 3-vectors, and the transformations of the spinors of one minimal left ideal of $Cl(1,3)$ into another minimal left ideal. This, among others, has implications for understanding the observed non-conservation of parity.[1]

4.1   Transformations of Clifford numbers

We will follow the approach [56–60] in which spinors are constructed in terms of nilpotents formed from the spacetime basis vectors represented as generators of the Clifford algebra

$$\gamma_a \cdot \gamma_b \equiv \tfrac{1}{2} \left(\gamma_a \gamma_b + \gamma_b \gamma_a \right) = \eta_{ab},$$
$$\gamma_a \wedge \gamma_b \equiv \tfrac{1}{2} \left(\gamma_a \gamma_b - \gamma_b \gamma_a \right). \tag{4.1}$$

The inner, symmetric, product of basis vectors gives the metric. The outer, antisymmetric, product of basis vectors gives a basis bivector.

---

[1]This chapter is based on Refs. [60, 65–67].

The generic Clifford number is

$$\Phi = \varphi^A \gamma_A, \tag{4.2}$$

where $\gamma_A \equiv \gamma_{a_1 a_2 \ldots a_r} \equiv \gamma_{a_1} \wedge \gamma_{a_2} \wedge \ldots \wedge \gamma_{a_r}, \quad r = 0, 1, 2, 3, 4$.

As we will see in the next sections, spinors are particular Clifford numbers, $\Psi = \psi^\alpha \xi_\alpha$, where ξ_α are spinor basis elements, composed from γ_A. Let us now consider transformation properties of Clifford numbers.

In general, a Clifford number transforms according to

$$\Phi \rightarrow \Phi' = R\,\Phi\,S. \tag{4.3}$$

Here R and S are Clifford numbers, e.g., $R = e^{\frac{1}{2}\alpha^A \gamma_A}$, $S = e^{\frac{1}{2}\beta^A \gamma_A}$.

In particular, if $S = 1$, we have

$$\Phi \rightarrow \Phi' = R\,\Phi. \tag{4.4}$$

As an example, let us consider the case

$$R = e^{\frac{1}{2}\alpha \gamma_1 \gamma_2} = \cos\frac{\alpha}{2} + \gamma_1 \gamma_2 \sin\frac{\alpha}{2}, \qquad S = e^{\frac{1}{2}\beta \gamma_1 \gamma_2} = \cos\frac{\beta}{2} + \gamma_1 \gamma_2 \sin\frac{\beta}{2} \tag{4.5}$$

and examine [66], how various Clifford numbers,

$$X = X^C \gamma_C, \tag{4.6}$$

transform under (4.3), which now reads:

$$X \rightarrow X' = R\,X\,S. \tag{4.7}$$

(i) If $X = X^1 \gamma_1 + X^2 \gamma_2$ then

$$X' = X^1 \left(\gamma_1 \cos\frac{\alpha - \beta}{2} + \gamma_2 \sin\frac{\alpha - \beta}{2} \right)$$

$$+ X^2 \left(-\gamma_1 \sin\frac{\alpha - \beta}{2} + \gamma_2 \cos\frac{\alpha - \beta}{2} \right). \tag{4.8}$$

(ii) If $X = X^3 \gamma_3 + X^{123} \gamma_{123}$ then

$$X' = X^3 \left(\gamma_3 \cos\frac{\alpha + \beta}{2} + \gamma_{123} \sin\frac{\alpha + \beta}{2} \right)$$

$$+ X^{123} \left(-\gamma_3 \sin\frac{\alpha + \beta}{2} + \gamma_{123} \cos\frac{\alpha + \beta}{2} \right). \tag{4.9}$$

(iii) If $X = s\underline{1} + X^{12} \gamma_{12}$, then

$$X' = s \left(\underline{1} \cos\frac{\alpha + \beta}{2} + \gamma_{12} \sin\frac{\alpha + \beta}{2} \right)$$

$$+X^{12}\left(-\underline{1}\sin\frac{\alpha+\beta}{2}+\gamma_{12}\cos\frac{\alpha+\beta}{2}\right). \tag{4.10}$$

(iv) If $X = \tilde{X}^1\gamma_5\gamma_1 + \tilde{X}^2\gamma_5\gamma_2$, then

$$X' = \tilde{X}^1\left(\gamma_5\gamma_1\cos\frac{\alpha-\beta}{2}+\gamma_5\gamma_2\sin\frac{\alpha-\beta}{2}\right)$$

$$+\tilde{X}^2\left(-\gamma_5\gamma_1\sin\frac{\alpha-\beta}{2}+\gamma_5\gamma_2\cos\frac{\alpha-\beta}{2}\right). \tag{4.11}$$

Usual rotations of vectors or pseudovectors are reproduced, if the angle β for the right transformation is equal to minus angle for the left transformation, i.e., if $\beta = -\alpha$. Then all other transformations which mix the grade vanish. But in general, if $\beta \neq \alpha$, the transformation (4.7) mixes the grade.

4.2 Clifford algebra and spinors in Minkowski space

Let us introduce a new basis, called the Witt basis,

$$\begin{aligned}\theta_1 &= \tfrac{1}{2}(\gamma_0+\gamma_3)\,, \theta_2 = \tfrac{1}{2}(\gamma_1+i\gamma_2)\,,\\ \bar{\theta}_1 &= \tfrac{1}{2}(\gamma_0-\gamma_3)\,, \bar{\theta}_2 = \tfrac{1}{2}(\gamma_1-i\gamma_2),\end{aligned} \tag{4.12}$$

where

$$\gamma_a = (\gamma_0, \gamma_1, \gamma_2, \gamma_3). \tag{4.13}$$

The new basis vectors satisfy

$$\{\theta_a, \bar{\theta}_b\} = \eta_{ab}, \quad \{\theta_a, \theta_b\} = 0, \quad \{\bar{\theta}_a, \bar{\theta}_b\} = 0, \tag{4.14}$$

which are fermionic anticommutation relations. We now observe that the product

$$f = \bar{\theta}_1\bar{\theta}_2 \tag{4.15}$$

satisfies

$$\bar{\theta}_a f = 0\,, \qquad a = 1, 2. \tag{4.16}$$

Here f can be interpreted a "vacuum", and $\bar{\theta}_a$ can be interpreted as operators that annihilate f.

An object constructed as a superposition

$$\Psi = (\psi^0 + \psi^1\theta_1 + \psi^2\theta_2 + \psi^{12}\theta_1\theta_2)f \tag{4.17}$$

is a 4-component spinor. It is convenient to change the notation:

$$\Psi = (\psi^1 + \psi^2\theta_1\theta_2 + \psi^3\theta_1 + \psi^4\theta_2)f = \psi^\alpha\xi_\alpha\,, \qquad \alpha = 1, 2, 3, 4 \tag{4.18}$$

where ξ_α is the spinor basis.

The even part of the above expression is a left handed spinor

$$\Psi_L = (\psi^1 + \psi^2\theta_1\theta_2)\,\bar\theta_1\bar\theta_2, \tag{4.19}$$

whereas the odd part is a right handed spinor

$$\Psi_R = (\psi^3\theta_1 + \psi^4\theta_2)\bar\theta_1\bar\theta_2. \tag{4.20}$$

We can verify that the following relations are satisfied:

$$i\gamma_5\Psi_L = -\Psi_L, \qquad i\gamma_5\Psi_R = \Psi_R. \tag{4.21}$$

Under the transformations

$$\Psi \to \Psi' = \mathrm{R}\Psi, \tag{4.22}$$

where

$$\mathrm{R} = \exp[\tfrac{1}{2}\gamma_{a_1}\gamma_{a_2}\varphi], \tag{4.23}$$

the Clifford number Ψ transforms as a spinor.

As an example let us consider the case

$$\mathrm{R} = e^{\frac{1}{2}\gamma_1\gamma_2\varphi} = \cos\frac{\varphi}{2} + \gamma_1\gamma_2\sin\frac{\varphi}{2}. \tag{4.24}$$

Then we have

$$\Psi \to \Psi' = \mathrm{R}\Psi = \left(e^{\frac{i\phi}{2}}\psi^1 + e^{-\frac{i\phi}{2}}\psi^2\theta_1\theta_2 + e^{\frac{i\phi}{2}}\psi^3\theta_1 + e^{-\frac{i\phi}{2}}\psi^4\theta_2 \right) f. \tag{4.25}$$

This is the well-known transformation of a 4-component spinor.

4.3   Four independent spinors

There exist four different possible vacua [60, 61, 68]:

$$f_1 = \bar\theta_1\bar\theta_2, \qquad f_2 = \theta_1\theta_2, \qquad f_3 = \theta_1\bar\theta_2, \qquad f_3 = \bar\theta_1\theta_2 \tag{4.26}$$

to which there correspond four different kinds of spinors:

$$\begin{aligned}
\Psi^1 &= (\psi^{11} + \psi^{21}\theta_1\theta_2 + \psi^{31}\theta_1 + \psi^{41}\theta_2)f_1 \\
\Psi^2 &= (\psi^{12} + \psi^{22}\bar\theta_1\bar\theta_2 + \psi^{32}\bar\theta_1 + \psi^{42}\bar\theta_2)f_2 \\
\Psi^3 &= (\psi^{13}\bar\theta_1 + \psi^{23}\theta_2 + \psi^{33} + \psi^{43}\bar\theta_1\theta_2)f_3 \\
\Psi^4 &= (\psi^{14}\theta_1 + \psi^{24}\bar\theta_2 + \psi^{34} + \psi^{44}\theta_1\bar\theta_2)f_4.
\end{aligned} \tag{4.27}$$

Each of those spinors lives in a different minimal left ideal of $Cl(1,3)$, or in general, of its complexified version if we assume complex $\psi^{\alpha i}$.

An arbitrary element of $Cl(1,3)$ is the sum:

$$\Phi = \Psi^1 + \Psi^2 + \Psi^3 + \Psi^4 = \psi^{\alpha i}\xi_{\alpha i} \equiv \psi^{\tilde A}\xi_{\tilde A}, \tag{4.28}$$

where

$$\xi_{\tilde{A}} \equiv \xi_{\alpha i} = \{f_1,\ \theta_1\theta_2 f_1, ..., \theta_1 f_4,\ \bar{\theta}_2 f_4,\ f_4,\ \bar{\theta}_1\theta_2 f_4\}, \qquad (4.29)$$

is a spinor basis of $Cl(1,3)$. Here Φ is a generalized spinor.

In matrix notation we have

$$\psi^{\alpha i} = \begin{pmatrix} \psi^{11} & \psi^{12} & \psi^{13} & \psi^{14} \\ \psi^{21} & \psi^{22} & \psi^{23} & \psi^{24} \\ \psi^{31} & \psi^{32} & \psi^{33} & \psi^{34} \\ \psi^{41} & \psi^{42} & \psi^{43} & \psi^{44} \end{pmatrix}, \quad \xi_{\tilde{A}} \equiv \xi_{\alpha i} = \begin{pmatrix} f_1 & f_2 & \bar{\theta}_1 f_3 & \theta_1 f_4 \\ \theta_1\theta_2 f_1 & \bar{\theta}_1\bar{\theta}_2 f_2 & \theta_2 f_3 & \bar{\theta}_2 f_4 \\ \theta_1 f_1 & \bar{\theta}_1 f_2 & f_3 & f_4 \\ \theta_2 f_1 & \bar{\theta}_2 f_2 & \bar{\theta}_1\theta_2 f_3 & \theta_1\bar{\theta}_2 f_4 \end{pmatrix}.$$

$$(4.30)$$

Here, for instance, the second column in the left matrix contains the components of the spinor of the second left ideal. Similarly, the second column in the right matrix contains the basis elements of the second left ideal.

A general transformation is

$$\Phi = \psi^{\tilde{A}}\xi_{\tilde{A}} \rightarrow \Phi' = R\,\Phi\,S = \psi^{\tilde{A}}\xi'_{\tilde{A}} = \psi^A L_{\tilde{A}}{}^{\tilde{B}}\xi_B = \psi'^{\tilde{B}}\xi_{\tilde{B}} \qquad (4.31)$$

where

$$\xi'_{\tilde{A}} = R\xi_{\tilde{A}}S = L_{\tilde{A}}{}^{\tilde{B}}\xi_{\tilde{B}}, \qquad \psi'^{\tilde{B}} = \psi^{\tilde{A}}L_{\tilde{A}}{}^{\tilde{B}}. \qquad (4.32)$$

This is an active transformation, because it changes an object Φ into another object Φ'.

The transformation from the left,

$$\Phi' = R\,\Phi, \qquad (4.33)$$

reshuffles the components within each left ideal, whereas the transformation from the right,

$$\Phi' = \Phi\,S, \qquad (4.34)$$

reshuffles the left ideals.

4.4 Behavior of spinors under Lorentz transformations

Let us consider the following transformation of the basis vectors

$$\gamma_a \rightarrow \gamma'_a = R\gamma_a R^{-1}, \qquad a = 0, 1, 2, 3, \qquad (4.35)$$

where R is a proper or improper Lorentz transformation. A generalized spinor, $\Phi \in Cl(1,3)$, composed of γ_a, then transforms according to

$$\Phi = \psi^{\tilde{A}}\xi_{\tilde{A}} \rightarrow \Phi' = \psi^{\tilde{A}}\xi'_{\tilde{A}} = \psi^A R\xi_B R^{-1} = R\,\Phi\,R^{-1}. \qquad (4.36)$$

The transformation (4.35) of the basis vectors has for a consequence that the object Φ does not transform only from the right, but also from the left.

This had led Piazzese to the conclusion that spinors cannot be interpreted as the minimal ideals of Clifford algebras [69].

But if the reference frame transforms as

$$\gamma_a \to \gamma_a' = \mathrm{R}\,\gamma_a, \tag{4.37}$$

then

$$\Phi = \psi^{\tilde{A}}\xi_{\tilde{A}} \to \Phi' = \psi^{\tilde{A}}\xi_{\tilde{A}}' = \psi^{\tilde{A}}\mathrm{R}\,\xi_{\tilde{B}} = \mathrm{R}\,\Phi. \tag{4.38}$$

This is a transformation of a spinor. Therefore, the description of spinors in terms of ideals is consistent.

As we have seen in Sec. 4.1, the transformation (4.37) is also a possible transformation within a Clifford algebra. It is a transformation that changes the grade of a basis element. Usually, we do not consider such transformations of basis vectors. Usually reference frames are "rotated" (Lorentz rotated) according to

$$\gamma_a \to \gamma_a' = \mathrm{R}\,\gamma_a\mathrm{R}^{-1} = L_a{}^b\gamma_b, \tag{4.39}$$

where $L_a{}^b$ is a proper or improper Lorentz transformation. Therefore, a "rotated" observer sees (generalized) spinors transformed according to

$$\Phi \to \Phi' = \mathrm{R}\,\Phi\,\mathrm{R}^{-1}. \tag{4.40}$$

With respect to a new reference frame, the object $\Phi = \psi^{\tilde{A}}\xi_{\tilde{A}}$ is expanded according to

$$\Phi = \psi'^{\tilde{A}}\xi_{\tilde{A}}', \tag{4.41}$$

where

$$\psi'^{\tilde{A}} = \psi^{\tilde{B}}(L^{-1})_{\tilde{B}}{}^{\tilde{A}}. \tag{4.42}$$

Recall that $\alpha, \beta = 1, 2, 3, 4$, and $i, j = 1, 2, 3, 4$. The corresponding matrix $\psi^{\alpha i}$ transforms from the left and from the right.

If the observer, together with the reference frame, starts to rotate, then after having exhibited the $\varphi = 2\pi$ turn, he observes the same spinor Ψ, as he did at $\varphi = 0$. The sign of the spinor did not change, because this was just a passive transformation, so that the same (unchanged) objects was observed from the transformed (rotated) references frames at different angles φ. In the new reference frame the object was observed to be transformed according to $\Psi' = R\Psi R^{-1}$. There must also exist the corresponding *active* transformation such that in a fixed reference frame the spinor transforms as $\Psi' = R\Psi R^{-1}$. In the following we will demonstrate on specific examples

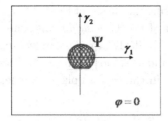
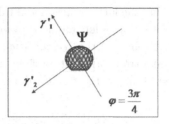
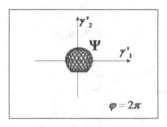

If the observer, together with the Reference frame, starts to rotate, Then after having exhibited the $\varphi = 2\pi$ turn, he observes the same spinor Ψ, as he did at $\varphi = 0$.

The sign of the spinor did not change.

Fig. 4.1 Illustration of a passive transformation, namely the rotation of the reference frame. In the new reference frame the object —for instance a spinor Ψ— is observed to be transformed according to $\Psi' = R\Psi R^{-1}$. There must also exist the corresponding active transformation, such that in a fixed reference frame the spinor Ψ transform as $\Psi' = R\Psi R^{-1}$.

how spinors behave under such transformations. First we will consider a rotation and next the space inversion.

4.4.1 *Rotation*

Let us consider the following rotation:

$$\gamma_0 \to \gamma_0, \quad \gamma_1 \to \gamma_1, \quad \gamma_2 \to \gamma_2 \cos \vartheta + \gamma_3 \sin \vartheta$$
$$\gamma_3 \to -\gamma_2 \sin \vartheta + \gamma_3 \cos \vartheta. \tag{4.43}$$

In the case $\vartheta = \pi$, we have

$$\gamma_0 \to \gamma_0, \quad \gamma_1 \to \gamma_1, \quad \gamma_2 \to -\gamma_2, \quad \gamma_3 \to -\gamma_3. \tag{4.44}$$

The Witt basis then transforms as

$$\theta_1 \to \bar{\theta}_1, \quad \theta_2 \to \bar{\theta}_2, \quad \bar{\theta}_1 \to \theta_1, \quad \bar{\theta}_2 \to \theta_2. \tag{4.45}$$

A consequence is that, e.g., a spinor of the first left ideal transforms as

$$(\psi^{11} + \psi^{21}\theta_1\theta_2 + \psi^{31}\theta_1 + \psi^{41}\theta_2)\,\bar{\theta}_1\bar{\theta}_2 \;\to\; (\psi^{11} + \psi^{21}\bar{\theta}_1\bar{\theta}_2 + \psi^{31}\bar{\theta}_1 + \psi^{41}\bar{\theta}_2)\,\theta_1\theta_2. \tag{4.46}$$

By inspecting the latter relation and taking into account Eqs. (4.27), (4.19), (4.20), we see that a left handed spinor of the *first ideal* transforms into a left handed spinor of the *second ideal*. Similarly, a right handed spinor of the first ideal transforms into a right handed spinor of the second ideal.

In general, under the $\vartheta = \pi$ rotation in the (γ_2, γ_3) plane, a generalized spinor

$$
\begin{aligned}
\Phi = {} & (\psi^{11} + \psi^{21}\theta_1\theta_2 + \psi^{31}\theta_1 + \psi^{41}\theta_2)\bar{\theta}_1\bar{\theta}_2 \\
& + (\psi^{12} + \psi^{22}\bar{\theta}_1\theta_2 + \psi^{32}\bar{\theta}_1 + \psi^{42}\theta_2)\theta_1\theta_2 \\
& + (\psi^{13}\bar{\theta}_1 + \psi^{23}\theta_2 + \psi^{33} + \psi^{43}\bar{\theta}_1\theta_2)\theta_1\theta_2 \\
& + (\psi^{14}\theta_1 + \psi^{24}\bar{\theta}_2 + \psi^{34} + \psi^{44}\theta_1\bar{\theta}_2)\bar{\theta}_1\theta_2
\end{aligned}
\tag{4.47}
$$

transforms into

$$
\begin{aligned}
\Phi' = {} & (\psi^{11} + \psi^{21}\bar{\theta}_1\bar{\theta}_2 + \psi^{31}\bar{\theta}_1 + \psi^{41}\bar{\theta}_2)\theta_1\theta_2 \\
& + (\psi^{12} + \psi^{22}\theta_1\theta_2 + \psi^{32}\theta_1 + \psi^{42}\theta_2)\bar{\theta}_1\bar{\theta}_2 \\
& + (\psi^{13}\theta_1 + \psi^{23}\bar{\theta}_2 + \psi^{33} + \psi^{43}\theta_1\bar{\theta}_2)\bar{\theta}_1\theta_2 \\
& + (\psi^{14}\bar{\theta}_1 + \psi^{24}\theta_2 + \psi^{34} + \psi^{44}\bar{\theta}_1\theta_2)\theta_1\bar{\theta}_2
\end{aligned}
\tag{4.48}
$$

The matrix of components

$$
\psi^{\alpha i} = \begin{pmatrix} \psi^{11} & \psi^{12} & \psi^{13} & \psi^{14} \\ \psi^{21} & \psi^{22} & \psi^{23} & \psi^{24} \\ \psi^{31} & \psi^{32} & \psi^{33} & \psi^{34} \\ \psi^{41} & \psi^{42} & \psi^{43} & \psi^{44} \end{pmatrix} \quad \text{transforms into} \quad \psi'^{\alpha i} = \begin{pmatrix} \psi^{12} & \psi^{11} & \psi^{14} & \psi^{13} \\ \psi^{22} & \psi^{21} & \psi^{24} & \psi^{23} \\ \psi^{32} & \psi^{31} & \psi^{34} & \psi^{33} \\ \psi^{42} & \psi^{41} & \psi^{44} & \psi^{43} \end{pmatrix}.
\tag{4.49}
$$

We see that in the transformed matrix, the first and the second column are interchanged. Similarly, also the third and forth column are interchanged. Different columns represent different left minimal ideals of $Cl(1,3)$, and thus different spinors.

Let us now focus our attention on the spinor basis states of the first and second ideal:

$$
\xi_{11} = \bar{\theta}_1\bar{\theta}_2, \quad \xi_{21} = \theta_1\theta_2\bar{\theta}_1\bar{\theta}_2, \quad \xi_{12} = \theta_1\theta_2, \quad \xi_{22} = \bar{\theta}_1\bar{\theta}_2\theta_1\theta_2, \tag{4.50}
$$

which span the left handed part of the 4-component spinor (see Eqs. (4.19), (4.20)).

Under the $\vartheta = \pi$ rotation (4.44), (4.45), we have

$$
\xi_{11} \to \xi_{12}, \quad \xi_{21} \to \xi_{22}, \quad \xi_{12} \to \xi_{11}, \quad \xi_{22} \to \xi_{21}, \tag{4.51}
$$

which means that the spin 1/2 state of the 1st ideal transforms into the spin state of the 2nd ideal, and vice versa. The above states are eigenvalues of the spin operator, $-\frac{i}{2}\gamma_1\gamma_2$,

$$
-\frac{i}{2}\gamma_1\gamma_2\,\xi_{11} = \frac{1}{2}\xi_{11}, \qquad -\frac{i}{2}\gamma_1\gamma_2\,\xi_{21} = -\frac{1}{2}\xi_{21}, \tag{4.52}
$$

$$-\frac{i}{2}\gamma_1\gamma_2\,\xi_{12} = -\frac{1}{2}\xi_{12}\,, \qquad -\frac{i}{2}\gamma_1\gamma_2\,\xi_{22} = \frac{1}{2}\xi_{22}\,. \tag{4.53}$$

Let us now introduce the new basis states

$$\begin{aligned}
\xi^1_{1/2} &= \tfrac{1}{\sqrt{2}}(\xi_{11} + \xi_{22})\,, \quad \xi^2_{1/2} = \tfrac{1}{\sqrt{2}}(\xi_{11} - \xi_{22})\,, \\
\xi^1_{-1/2} &= \tfrac{1}{\sqrt{2}}(\xi_{21} + \xi_{12})\,, \quad \xi^2_{-1/2} = \tfrac{1}{\sqrt{2}}(\xi_{21} - \xi_{12})\,,
\end{aligned} \tag{4.54}$$

which are superpositions of the states of the 1st and the 2nd ideal. Under the rotation (4.44),(4.45) we have

$$\begin{aligned}
\xi^1_{1/2} &\to \tfrac{1}{\sqrt{2}}(\xi_{12} + \xi_{21}) = \xi^1_{-1/2}\,, \\
\xi^1_{-1/2} &\to \tfrac{1}{\sqrt{2}}(\xi_{22} + \xi_{11}) = \xi^1_{1/2}\,,
\end{aligned} \tag{4.55}$$

$$\begin{aligned}
\xi^2_{1/2} &\to \tfrac{1}{\sqrt{2}}(\xi_{12} - \xi_{21}) = -\xi^2_{-1/2}\,, \\
\xi^2_{-1/2} &\to \tfrac{1}{\sqrt{2}}(\xi_{22} - \xi_{11}) = -\xi^2_{1/2}\,.
\end{aligned} \tag{4.56}$$

These states also have definite spin projection:

$$-\frac{i}{2}\gamma_1\gamma_2\xi^1_{\pm1/2} = \pm\frac{1}{2}\xi^1_{\pm1/2}\,, \tag{4.57}$$

$$-\frac{i}{2}\gamma_1\gamma_2\xi^2_{\pm1/2} = \pm\frac{1}{2}\xi^2_{\pm1/2}\,. \tag{4.58}$$

The states (4.55) have the property that under the $\vartheta = \pi$ rotation, the spin $1/2$ state $\xi^1_{1/2}$ transforms into the spin $-1/2$ state $\xi^1_{-1/2}$, and vice versa. Analogous hold for the other set of states, $\xi^2_{1/2}$, $\xi^2_{-1/2}$.

Let us stress again that the transformation in the above example is of the type $\Phi' = R\Phi R^{-1}$. This is a reason that, under such a transformation, a spinor of one ideal is transformed into the spinor of a different ideal. A transformation R^{-1}, acting from the right, mixes the ideals. Another kind of transformation is $\Phi' = R\Phi$, in which case there is no mixing of ideals. Such are the usual transformations of spinors. By considering the objects of the entire Clifford algebra and possible transformations among them, we find out that spinors are not a sort of objects that transform differently than vectors under rotations. They can transform under rotations in the same way as vectors, i.e., according to $\Phi' = R\Phi R^{-1}$. Here Φ can be a vector, spinor or any other object of Clifford algebra. In addition to this kind of transformations, there exist also the other kind of transformations, namely, $\Phi' = R\Phi$, where again Φ can be any object of $Cl(1,3)$, including a vector or a spinor. These are particular cases of the more general transformations, $\Phi' = R\Phi S$, considered in Sec. 4.1.

4.4.2 Space inversion

Let us now consider space inversion, under which the basis vectors of a reference frame transform according to

$$\gamma_0 \to \gamma_0' = \gamma_0 \,, \quad \gamma_r \to \gamma_r' = -\gamma_r \,, \quad r = 1, 2, 3 \,. \tag{4.59}$$

The vectors of the Witt basis (4.12) then transform as

$$
\begin{aligned}
\theta_1 &\to \tfrac{1}{2}(\gamma_0 - \gamma_3) = \bar{\theta}_1 \,, \\
\theta_2 &\to \tfrac{1}{2}(-\gamma_1 - i\gamma_2) = -\theta_2, \\
\bar{\theta}_1 &\to \tfrac{1}{2}(\gamma_0 + \gamma_3) = \theta_1 \,, \\
\bar{\theta}_2 &\to \tfrac{1}{2}(-\gamma_1 + i\gamma_2) = -\bar{\theta}_2 \,.
\end{aligned} \tag{4.60}
$$

A spinor of the first left ideal transforms as [60]

$$(\psi^{11}\underline{1} + \psi^{21}\theta_1\theta_2 + \psi^{31}\theta_1 + \psi^{41}\theta_2)\,\bar{\theta}_1\bar{\theta}_2$$

$$\to (-\psi^{11}\underline{1} + \psi^{21}\bar{\theta}_1\theta_2 - \psi^{31}\bar{\theta}_1 + \psi^{41}\theta_2)\,\theta_1\bar{\theta}_2 \,. \tag{4.61}$$

The latter equation shows that a left handed spinor of the first ideal transforms into a right handed spinor of the third ideal.

In general, under space inversion, the matrix of the spinor basis elements

$$
\xi_{\alpha i} = \begin{pmatrix}
f_1 & f_2 & \bar{\theta}_1 f_3 & \theta_1 f_4 \\
\theta_1 \theta_2 f_1 & \bar{\theta}_1 \bar{\theta}_2 f_2 & \theta_2 f_3 & \bar{\theta}_2 f_4 \\
\theta_1 f_1 & \bar{\theta}_1 f_2 & f_3 & f_4 \\
\theta_2 f_1 & \bar{\theta}_2 f_2 & \bar{\theta}_1 \theta_2 f_3 & \theta_1 \bar{\theta}_2 f_4
\end{pmatrix}, \tag{4.62}
$$

transforms into

$$
\xi_{\alpha i}' = \begin{pmatrix}
-f_3 & -f_4 & -\theta_1 f_1 & -\bar{\theta}_1 f_2 \\
\bar{\theta}_1 \theta_2 f_3 & \theta_1 \bar{\theta}_2 f_4 & \theta_2 f_1 & \bar{\theta}_2 f_2 \\
-\bar{\theta}_1 f_3 & -\theta_1 f_4 & -f_1 & -f_2 \\
\theta_2 f_3 & \bar{\theta}_2 f_4 & \theta_1 \theta_2 f_1 & \bar{\theta}_1 \bar{\theta}_2 f_2
\end{pmatrix}. \tag{4.63}
$$

The matrix of components

$$
\psi^{\alpha i} = \begin{pmatrix}
\psi^{11} & \psi^{12} & \psi^{13} & \psi^{14} \\
\psi^{21} & \psi^{22} & \psi^{23} & \psi^{24} \\
\psi^{31} & \psi^{32} & \psi^{33} & \psi^{34} \\
\psi^{41} & \psi^{42} & \psi^{43} & \psi^{44}
\end{pmatrix}
$$

transform into $\psi^{\alpha i} = \begin{pmatrix}
-\psi^{33} & -\psi^{34} & -\psi^{31} & -\psi^{32} \\
\psi^{43} & \psi^{44} & \psi^{41} & \psi^{42} \\
-\psi^{13} & -\psi^{14} & -\psi^{11} & -\psi^{12} \\
\psi^{23} & \psi^{24} & \psi^{21} & \psi^{22}
\end{pmatrix}.$ (4.64)

By comparing (4.62) and (4.63), or by inspecting (4.64), we find that the spinor of the 1st ideal transforms into the spinor of the 3rd ideal, and the spinor of the 2nd ideal transforms into the spinor of the 4th ideal.

4.5   Generalized Dirac equation (Dirac-Kähler equation)

Let us now consider the Clifford algebra valued fields, $\Phi(x)$, that depend on position $x \equiv x^\mu$ in spacetime. We will assume that a field Φ satisfies the following equation [62] (see also Refs. [60, 68]):

$$(i\gamma^\mu \partial_\mu - m)\Phi = 0, \qquad \Phi = \phi^A \gamma_A = \psi^{\tilde{A}}\xi_{\tilde{A}} = \psi^{\alpha i}\xi_{\alpha i}, \qquad (4.65)$$

where γ_A is a multivector basis of $Cl(1,3)$, and $\xi_{\tilde{A}} \equiv \xi_{\alpha i}$ is a spinor basis of $Cl(1,3)$, or more precisely, of its complexified version if $\psi^{\alpha i}$ are complex-valued. Here α is the spinor index of a left minimal ideal, whereas the i runs over four left ideals of $Cl(1,3)$.

Multiplying Eq. (4.65) from the left by $(\xi^{\tilde{A}})^\ddagger$, where $\ddagger$ is the operation of reversion that reverses the order of vectors in a product, and using the relation

$$\langle (\xi^{\tilde{A}})^\ddagger \gamma^\mu \xi_{\tilde{B}} \rangle_S \equiv (\gamma^\mu)^{\tilde{A}}{}_{\tilde{B}}, \qquad (4.66)$$

and where $\langle\ \rangle_S$ is the (properly normalized [71]) scalar part of an expression, we obtain the following matrix form of the equation (4.65):

$$\left(i(\gamma^\mu)^{\tilde{A}}{}_{\tilde{B}}\,\partial_\mu - m\,\delta^{\tilde{A}}{}_{\tilde{B}}\right)\psi^{\tilde{B}} = 0. \qquad (4.67)$$

The 16×16 matrices can be factorized according to

$$(\gamma^\mu)^{\tilde{A}}{}_{\tilde{B}} = (\gamma^\mu)^\alpha{}_\beta\,\delta^i{}_j, \qquad (4.68)$$

where $(\gamma^\mu)^\alpha{}_\beta$ are 4×4 Dirac matrices. Using the latter relation (4.68), we can write Eq. (4.67) as

$$\left(i(\gamma^\mu)^\alpha{}_\beta\,\partial_\mu - m\,\delta^\alpha{}_\beta\right)\psi^{\beta i} = 0, \qquad (4.69)$$

or more simply,

$$(i\gamma^\mu \partial_\mu - m)\psi^i = 0. \qquad (4.70)$$

In the last equation we have omitted the spinor index α.

The action that leads to the generalized Dirac equation (4.65) is

$$I = \int \mathrm{d}^4x\ \bar{\psi}^i(i\gamma^\mu \partial_\mu - m)\psi^j z_{ij}. \qquad (4.71)$$

This is an action that describes four spinors ψ^i, belonging to the four minimal left ideals of $Cl(1,3)$. Here z_{ij} is the metric in the space of ideals. It is a part of the metric

$$(\xi_{\tilde{A}})^\ddagger * \xi_{\tilde{B}} = z_{\tilde{A}\tilde{B}} = z_{(\alpha i)(\beta j)} = z_{\alpha\beta}z_{ij} \qquad (4.72)$$

of the Clifford algebra $Cl(1,3)$, represented in the basis $\xi_{\tilde{A}}$:

$$z_{ij} = \begin{pmatrix} 1 & 0 & 0 & 0 \\ 0 & 1 & 0 & 0 \\ 0 & 0 & -1 & 0 \\ 0 & 0 & 0 & -1 \end{pmatrix}, \qquad z_{\alpha\beta} = \begin{pmatrix} 1 & 0 & 0 & 0 \\ 0 & 1 & 0 & 0 \\ 0 & 0 & -1 & 0 \\ 0 & 0 & 0 & -1 \end{pmatrix}. \qquad (4.73)$$

Gauge covariant action is

$$I = \int \mathrm{d}^4 x \; \bar{\psi}^i (i\gamma^\mu D_\mu - m)\psi^j z_{ij} , \qquad D_\mu \psi^i = \partial_\mu \psi^i + G_\mu{}^i{}_j \psi^j . \qquad (4.74)$$

This action contains the ordinary particles and mirror particles. The first and the second columns of the matrix $\psi^{\alpha i}$, written explicitly in Eq. (4.30) describe the ordinary particles, whereas the third and the forth column in (4.30) describe *mirror particles.*

The SU(2) gauge group acting within the 1st and 2nd ideal can be interpreted as the weak interaction gauge group for ordinary particles. The SU(2) gauge group acting within the 3rd and 4th ideal can be interpreted as the weak interaction gauge group for mirror particles. The corresponding two kinds of weak interaction gauge fields that can be transformed into each other by space inversion are contained in $G_\mu{}^i{}_j$, which is a generalized gauge field occurring in the covariant action (4.74).

Mirror particles were first proposed by Lee and Yang [72]. Subsequently, the idea of mirror particles has been pursued by Kobzarev et al. [73], and in Refs. [74,75,78–81]. The possibility that mirror particles are responsible for dark matter has been explored in many works, e.g., in [82–89]. For a review see [90]. A demonstration that mirror particles can be explained in terms of algebraic spinors (elements of Clifford algebras) was presented in Ref. [60].

Chapter 5

Quantum Fields as Basis Vectors

It is shown that the generators of Clifford algebras behave as creation and annihilation operators for fermions and bosons. They can create extended objects, such as strings and branes, and can induce curved metric of our spacetime. At a fixed point, we consider the Clifford algebra $Cl(8)$ of the 8-dimensional phase space, and show that one quarter of the basis elements of $Cl(8)$ can represent all known particles of the first generation of the Standard Model, whereas the other three quarters are invisible to us and can thus correspond to dark matter. This chapter is a shortened and modified version of Ref. [91].

5.1 Introduction

Quantization of a classical theory is a procedure that appears somewhat enigmatic. It is not a derivation in a mathematical sense. It is a recipe of how to replace, e.g., the classical phase space variables, satisfying the Poisson bracket relations, with the operators satisfying the corresponding commutation relations [92]. What is a deeper meaning for replacement is usually not explained, only that it works. A quantized theory so obtained does work and successfully describes the experimental observations of quantum phenomena.

On the other hand, there exists a very useful tool for description of geometry of a space of arbitrary dimension and signature [93–99, 102–106]. This is Clifford algebra. Its generators are the elements that satisfy the well-known relations, namely that the anticommutators of two generators are proportional to the components of a symmetric metric tensor. The space spanned by those generators is a vector space. It can correspond to a physical space, for instance to our usual three dimensional space, or to the four dimensional spacetime. The generators of a Clifford algebra are thus basis vectors of a physical space. We will interpret this as a space of all possible positions that the *center of mass* of a physical object can posses. A physical object has an extension that can be described by an effective oriented area, volume, etc. While the center of mass position is described by a vector, the oriented area is described by a *bivector*, the oriented volume by a trivector, etc. In general, an extended object is described [49,99,104–106] by a superposition of scalars, vectors, bivectors, trivectors, etc., i.e., by an element of the Clifford algebra. The Clifford algebra associated with an extended object is a space, called *Clifford space*[1].

Besides the Clifford algebras whose generators satisfy the anticommutation relations, there are also the algebras whose generators satisfy commutation relations, such that the commutators of two generators are equal to the components of a metric, which is now *antisymmetric*. The Clifford algebras with a symmetric metric are called *orthogonal Clifford algebras*, whereas the Clifford algebras with an antisymmetric metric are called *symplectic Clifford algebras* [107].

We will see that symplectic basis vectors are in fact quantum mechanical operators of bosons [68,108]. The Poisson brackets of two classical coordinates are *equal* to the commutators of two operators. This is so, because the Poisson bracket consists of the derivative and the symplectic metric which is equal to the commutator of two symplectic basis vectors. The derivative acting on phase space coordinates yields the Kronecker delta and thus eliminates them from the expression. What remains is the commutator of the basis vectors.

Similarly, the basis vectors of an orthogonal Clifford algebra are quantum mechanical operators for fermions. This becomes evident in the new basis, the so called Witt basis. By using the latter basis vectors and their products, one can construct spinors.

Orthogonal and symplectic Clifford algebras can be extended to infinite dimensional spaces [68,108]. The generators of those infinite dimensional Clifford algebras are fermionic and bosonic field operators. In the case of fermions, a possible vacuum state can be the product of an infinite sequence of the operators [68,108]. If we act on such a vacuum with an operator that does not belong to the set of operators forming that vacuum,

---

[1]Here we did not go into the mathematical subtleties that become acute when the Clifford space is not flat but curved. Then, strictly speaking, the Clifford space is a *manifold*, such that the tangent space in any of its points is a Clifford algebra. If Clifford space is *flat*, then it is isomorphic to a Clifford algebra.

we obtain a "hole" in the vacuum. This hole behaves as a particle. The concept of the Dirac sea, which is today considered as obsolete, is revived within the field theories based on Clifford algebras. But in the latter theories we do not have only one vacuum, but many possible vacuums. This brings new possibilities for further development of quantum field theories and grand unification. Because the generators of Clifford algebras are basis vectors on the one hand, and field operators on the other hand, this opens a bridge towards quantum gravity. Namely, the expectation values of the "flat space" operators with respect to suitable quantum states composed of many fermions or bosons, can give "curved space" vectors, tangent to a manifold with non vanishing curvature. This observation paves the road to quantum gravity.

5.2 Clifford space as an extension of spacetime

Let us consider a *flat* space M whose points are *possible* positions of the center of mass P of a physical object $\mathcal{O}$. If the object's size is small in comparison to the distances to surrounding objects, then we can approximate the object with a point particle. The squared distance between two possible positions, with coordinates x^μ and $x^\mu + \Delta x^\mu$, is

$$\Delta s^2 = \Delta x^\mu g_{\mu\nu} \Delta x^\nu. \tag{5.1}$$

Here index μ runs over dimensions of the space M, and $g_{\mu\nu}$ is the metric tensor. For instance, in the case in which M is spacetime, $\mu = 0, 1, 2, 3$, and $g_{\mu\nu} = \eta_{\mu\nu} = \text{diag}(1, -1, -1, -1)$ is the Minkowski metric. The object $\mathcal{O}$ is then assumed to be extended in spacetime, i.e., to have an extension in a 3D space and in the direction x^0 that we call "time".

There are two possible ways of taking the square root of Δs^2.
Case I.

$$\Delta s = \sqrt{\Delta x^\mu g_{\mu\nu} \Delta x^\nu} \tag{5.2}$$

Case II.

$$\Delta x = \Delta x^\mu \gamma_\mu \tag{5.3}$$

In *Case* I, the square root is a *scalar*, i.e., the distance Δs.

In *Case* II, the square root is a *vector* Δx, expanded in term of the basis vectors γ_μ, satisfying the relations

$$\gamma_\mu \cdot \gamma_\nu \equiv \frac{1}{2}(\gamma_\mu \gamma_\nu + \gamma_\nu \gamma_\mu) = g_{\mu\nu}. \tag{5.4}$$

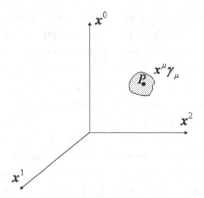

Fig. 5.1   The center of mass point P of an extended object $\mathcal{O}$ is described by a vector $x^\mu \gamma_\mu$.

If we write $\Delta x = \Delta x^\mu \gamma_\mu = (x^\mu - x_0^\mu)\gamma_\mu$ and take $x_0^\mu = 0$, we obtain $x = x^\mu \gamma_\mu$, which is the position vector of the object's $\mathcal{O}$ center of mass point P (Fig. 5.1), with x^μ being the coordinates of the point P.

In spite of being extended in spacetime and having many (practically infinitely many) degrees of freedom, we can describe our object $\mathcal{O}$ by only four coordinates x^μ, the components of a vector $x = x^\mu \gamma_\mu$.

The γ_μ satisfying the anticommutation relations (5.4) are generators of the Clifford algebra $Cl(1,3)$. A generic element of $Cl(1,3)$ is a superposition

$$X = \sigma \underline{1} + x^\mu \gamma_\mu + \frac{1}{2!} x^{\mu\nu} \gamma_\mu \wedge \gamma_\nu + \frac{1}{3!} x^{\mu\nu\rho} \gamma_\mu \wedge \gamma_\nu \wedge \gamma_\rho + \frac{1}{4!} x^{\mu\nu\rho\sigma} \gamma_\mu \wedge \gamma_\nu \wedge \gamma_\rho \wedge \gamma_\sigma,$$

$$(5.5)$$

where $\gamma_\mu \wedge \gamma_\nu$, $\gamma_\mu \wedge \gamma_\nu \wedge \gamma_\rho$ and $\gamma_\mu \wedge \gamma_\nu \wedge \gamma_\rho \wedge \gamma_\sigma$ are the antisymmetrized products $\gamma_\mu \gamma_\nu$, $\gamma_\mu \gamma_\nu \gamma_\rho$, and $\gamma_\mu \gamma_\nu \gamma_\rho \gamma_\sigma$, respectively. They represent basis bivectors, 3-vectors and 4-vectors, respectively. The terms in Eq. (5.5) describe a scalar, an oriented line, area, 3-volume and 4-volume. The antisymmetrized product of five gammas vanishes identically in four dimensions.

A question now arises as to whether the object X of Eq. (5.5) can describe an extended object in spacetime M_4. We have seen that $x = x^\mu \gamma_\mu$ describes the centre of mass position. We anticipate that $\frac{1}{2!} x^{\mu\nu} \gamma_\mu \wedge \gamma_\nu$ describes an oriented area associated with the extended object. Suppose that our object $\mathcal{O}$ is a closed string. At first approximation its is described just by its center of mass coordinates (Fig. 5.2a). At a better approximation it is described by the quantities $x^{\mu\nu}$, which are the projections of the oriented *area*, enclosed by the string, onto the coordinate planes (Fig. 5.2b). If we probe the string at a better resolution, we might find that it is not exactly a string, but a closed membrane (Fig. 5.3). The oriented volume, enclosed by this 2-dimensional membrane is described by the quantities $X^{\mu\nu\rho}$. At even better resolution we could eventually see that our object $\mathcal{O}$ is in fact a closed 3-dimensional membrane, enclosing a 4-volume, described by $x^{\mu\nu\rho\sigma}$. Our object $\mathcal{O}$ has *finite extension* in the 4-dimensional spacetime. It is like an instanton.

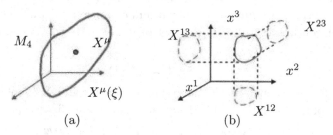

(a)   (b)

Fig. 5.2   With a closed string one can associate the center of mass coordinates (a), and the area coordinates (b).

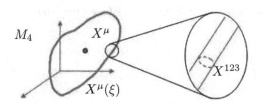

Fig. 5.3   Looking with a sufficient resolution one can detect eventual presence of volume degrees of freedom.

Let us now introduce a more compact notation by writing

$$X = \sum_{r=0}^{4} x^{\mu_1\mu_2\ldots\mu_r} \gamma_{\mu_1\mu_2\ldots\mu_r} \equiv x^M \gamma_M, \qquad (5.6)$$

where $\gamma_{\mu_1\mu_2\ldots\mu_r} \equiv \gamma_{\mu_1} \wedge \gamma_{\mu_2} \wedge \ldots \wedge \gamma_{\mu_r}$, and where we now assume $\mu_1 < \mu_2 < \ldots < \mu_r$, so that we do not need a factor $1/r!$. Here x^M are interpreted as quantities that describe an extended instantonic object in M_4. On the other hand, x^M are coordinates of a *point* in the 16-dimensional space, called *Clifford space* C. In other words, from the point of view of C, x^M describe a point in C.

The coordinates x^M of Clifford space can describe not only closed, but also open branes. For instance, a vector $x^\mu \gamma_\mu$ can denote position of a point event with respect to the origin (Fig. 5.1), or it can describe a string-like extended object (an instantonic string in spacetime). Similarly, a bivector $x^{\mu\nu} \gamma_\mu \wedge \gamma_\nu$ can describe a closed string (5.2a), or it can describe an open membrane. Whether the coordinates $x^M \equiv x^{\mu_1 \mu_2 \cdots \mu_r}$ describe a closed r-brane or an open $(r+1)$-brane is determined by the value of the scalar and pseudoscalar coordinates, i.e., by σ and $\tilde{\sigma}$ (for more details see Ref. [109]).

A continuous 1-dimensional set of points in C is a curve, a *worldline*, described by the mapping

$$x^M = X^M(\tau), \tag{5.7}$$

where τ is a monotonically increasing parameter and X^M embedding functions of the worldline in C. We assume that it satisfies the action principle

$$I[X^M] = \mathcal{M} \int d\tau \, (G_{MN} \dot{X}^M \dot{X}^N)^{1/2}, \tag{5.8}$$

where G_{MN} is the metric in C, and $\mathcal{M}$ a constant, analogous to mass. From the point of view of spacetime, the functions $X^M(\tau) \equiv X^{\mu_1 \mu_2 \cdots \mu_r}(\tau)$, $r = 0, 1, 2, 3, 4$, describe evolution of an extended instantonic object in spacetime. Some examples are in Fig. 5.4 (see also [104]).

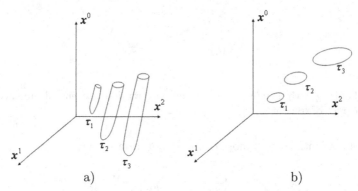

Fig. 5.4 Evolution of an instantonic cigar like (a) and a ring like (b) extended object in spacetime. At different values of τ, (e.g., at $\tau = \tau_1, \tau_2, \tau_3$), we have different extended instantonic objects that correspond to different 4D slices through Clifford space.

In this setup, there is no "block universe" in spacetime. There do not exist infinitely long worldlines or worldtubes in spacetime. Infinitely long worldlines exist in C-space, and in this sense a block universe exists in C-space.

The action (5.8) is invariant under reparameterizations of τ. A consequence is the constraint among the canonical momenta $P_M = \partial L/\partial \dot{X}^M = \mathcal{M}\dot{X}_M/\sqrt{g_{JK}\dot{X}^J\dot{X}^K}$:

$$P_M P^M - \mathcal{M}^2 = 0, \qquad (5.9)$$

where $P_M P^M = \eta_{MN} P^M P^N$. The metric of Clifford space is given by the scalar product of two basis vectors,

$$\eta_{MN} = \gamma_M^{\ddagger} * \gamma = <\gamma_M^{\ddagger} \gamma_N>_0, \qquad (5.10)$$

where "$\ddagger$" is the operation that reverses the order of vectors in the product $\gamma_M = \gamma_{\mu_1}\gamma_{\mu_2}...\gamma_{\mu_r}$, so that $\gamma_M^{\ddagger} = \gamma_{\mu_r}...\gamma_{\mu_2}\gamma_{\mu_1}$. The superscript "0" denotes the scalar part of an expression. For instance,

$$<\gamma_\mu\gamma_\nu>_0 = \eta_{\mu\nu}, \quad <\gamma_\mu\gamma_\nu\gamma_\alpha>_0 = 0, \quad <\gamma_\mu\gamma_\nu\gamma_\alpha\gamma_\beta>_0 = \eta_{\mu\beta}\eta_{\nu\alpha} - \eta_{\mu\alpha}\eta_{\nu\beta}. \qquad (5.11)$$

So we obtain

$$\eta_{MN} = \text{diag}(1,1,1,1,1,1,1,1,-1,-1,-1,-1,-1,-1,-1,-1), \qquad (5.12)$$

which means that the signature of C-sapce is $(++++++++----$ $----)$, or in short, $(8,8)$.

The quadratic form reads

$$\begin{aligned}
X^{\ddagger} * X &= \eta_{MN} x^M x^N \\
&= \sigma^2 + \eta_{\mu\nu}x^\mu x^\nu + (\eta_{\mu\beta}\eta_{\nu\alpha} - \eta_{\mu\alpha}\eta_{\nu\beta})x^{\mu\alpha}x^{\nu\beta} + \eta_{\mu\nu}\tilde{x}^\mu\tilde{x}^\nu - \tilde{\sigma}^2 \\
&= \eta_{\hat{\mu}\hat{\nu}}x^{\hat{\mu}}x^{\hat{\nu}} + \sigma^2 - \tilde{\sigma}^2, \qquad (5.13)
\end{aligned}$$

where $x^{\hat{\mu}} = (x^\mu, x^{\mu\nu}, \tilde{x}^\mu)$, with $\tilde{x}^\mu \equiv \frac{1}{3!}\epsilon^\mu_{\ \nu\rho\sigma}x^{\nu\rho\sigma}$ being the pseudoscalar coordinates, whereas σ is the scalar and $\tilde{\sigma} \equiv \frac{1}{4!}\epsilon_{\mu\nu\rho\sigma}x^{\mu\nu\rho\sigma}$ the pseudoscalar coordinate in C-space.

Upon quantization, P_M become operators $P_M = -i\partial/\partial x^M$, and the constraint (5.9) becomes the Klein-Gordon equation in C-space:

$$(\partial_M \partial^M + \mathcal{M}^2)\Psi(x^M) = 0. \qquad (5.14)$$

In the new coordinates,

$$s = \frac{1}{2}(\sigma + \tilde{\sigma}) \qquad \lambda = \frac{1}{2}(\sigma - \tilde{\sigma}), \qquad (5.15)$$

in which the quadratic form is

$$X^{\ddagger} * X = \eta_{\hat{\mu}\hat{\nu}}x^{\hat{\mu}}x^{\hat{\nu}} - 2s\lambda, \qquad (5.16)$$

the Klein-Gordon equation reads

$$\eta^{\hat{\mu}\hat{\mu}}\partial_{\hat{\mu}}\partial_{\hat{\nu}}\phi - 2\partial_s\partial_\lambda\phi = 0. \qquad (5.17)$$

If we take the ansatz

$$\phi(x^{\hat{\mu}}, s, \lambda) = e^{i\Lambda\lambda}\psi(s, x^{\hat{\mu}}), \tag{5.18}$$

then Eq. (5.17) becomes [110]

$$\eta^{\hat{\mu}\hat{\mu}}\partial_{\hat{\mu}}\partial_{\hat{\nu}}\phi - 2i\Lambda\partial_s\phi = 0, \tag{5.19}$$

i.e.,

$$i\frac{\partial\psi}{\partial s} = \frac{1}{2\Lambda}\eta^{\hat{\mu}\hat{\mu}}\partial_{\hat{\mu}}\partial_{\hat{\nu}}\psi. \tag{5.20}$$

This is the generalized *Stueckelberg equation*. It is like the Schrödinger equation, but it describes the evolution of the wave function $\psi(s, x^{\hat{\mu}})$ in the 14-dimensional space whose points are described by coordinates $x^{\hat{\mu}}$. The evolution parameter is s.

A remarkable feature of this setup is that the evolution parameter has a clear physical meaning: it is given in terms of the scalar, σ, and the pseudoscalar, $\tilde{\sigma}$, coordinate according to Eq. (5.15). The latter quantities, as shown before, are given by a configuration of the object, sampled in terms of the coordinates X^M of the Clifford space C.

The wave function $\psi(s, x^{\hat{\mu}})$ is the probability amplitude that at a given value of the evolution parameter s we will find an instantonic extended object with coordinates $x^{\hat{\mu}}$.

This is illustrated in Fig. 5.5. In principle all points of C-space are possible in the sense that we can find there an instantonic extended object . A wave packet determines a subset of point of C that are more probable to "host" the occurrence of an instantonic object (an event in C). The wave function determines the probability amplitude over the points of C. Its square determines the probability density. From the point of view of spacetime, wave function determines which instantonic extended objects are more likely to occur. It determines the probability amplitude, and its square determines the probability density of occurrence of a given instantonic extended object. The probability amplitude ψ is different at different values of the evolution parameter s. In other words, ψ changes (evolves) with s.

Instead of one extended object, described by x^M, we can consider several or many extended objects, described by x^{iM}, $i = 1, 2, ..., n$. They form an instantonic configuration $\{\mathcal{O}^i\} = \{\mathcal{O}^i\}, i = 1, 2, ..., n$. The space of all possible instantonic configurations will be called configuration space $\mathcal{C}$. The infinitesimal distance between two configurations, i.e., between two points in $\mathcal{C}$, is

$$\mathrm{d}S^2 = \eta_{(iM)(jN)}\mathrm{d}x^{iM}\mathrm{d}x^{jN}, \tag{5.21}$$

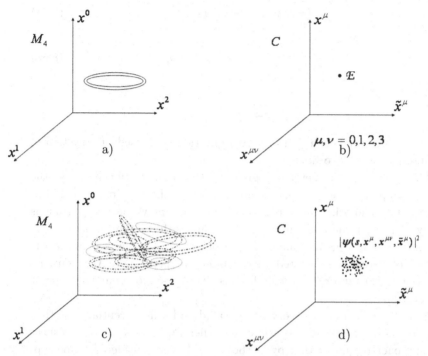

Fig. 5.5 Extended instantonic object in spacetime (a) is represented by a point in C-space (b). Quantum mechanically, the extended object is blurred (c). In C-space, we have a blurred point, i.e., a "cloud" of points occurring with probability density $|\psi(s, x^{\hat{\mu}})|^2$.

where $\eta_{(iM)(jN)} = \delta_{ij}\eta_{MN}$ is the metric of a *flat* configuration space.

We will assume that the Klein-Gordon equation (5.14) can be generalized so to hold for the wave function $\psi(x^{iM})$ in the space of instantonic configurations $\{\mathcal{O}^i\}$:

$$\left(\eta^{(iM)(jN)}\partial_{iM}\partial_{jN} + \mathcal{K}^2\right)\phi(x^{iM}) = 0 , \qquad \partial_{iM} \equiv \frac{\partial}{\partial X^{iM}}. \qquad (5.22)$$

Let us choose a particular extended object, $\mathcal{O}^1$, with coordinates $x^{1M} \equiv x^M = (\sigma, x^\mu, x^{\mu\nu}, \tilde{x}^\mu, \tilde{\sigma})$. The coordinates of the remaining extended objects within the configuration are $x^{2M}, x^{3M}, \dots$. Let us denote them $x^{\bar{i}M}, \bar{i} = 2, 3, \dots, N$. Following the same procedure as in Eqs. (5.15)–(5.20), we define s and λ according to (5.15) to the first object. We have thus split the coordinates x^{iM} of the configuration according to

$$x^{iM} = (s, \lambda, x^{\hat{\mu}}, x^{\bar{i}M}) = (s, \lambda, x^{\bar{M}}), \qquad (5.23)$$

where $x^{\bar{M}} = (x^{\hat{\mu}}, x^{\bar{i}M})$. By taking the ansatz

$$\phi(s, \lambda, x^{\bar{M}}) = e^{i\Lambda\lambda}\psi(s, x^{\bar{M}}), \tag{5.24}$$

Eq. (5.22) becomes

$$\eta^{\bar{M}\bar{N}}\partial_{\bar{M}}\partial_{\bar{N}}\psi - 2i\Lambda\partial_s\psi = 0, \tag{5.25}$$

i.e.,

$$i\frac{\partial}{\partial s}\psi = \frac{1}{2\Lambda}\eta^{\bar{M}\bar{N}}\partial_{\bar{M}}\partial_{\bar{N}}\psi. \tag{5.26}$$

Eq. (5.26) describes evolution of a configuration composed of a system of instantonic extended objects.

The evolution parameter s is given by the configuration itself (in the above example by one of its parts), and it distinguishes one instantonic configuration from another instantonic configuration. So we have a continuous family of instantonic configurations, evolving with s. Here, "instantonic configuration" or "instantonic extended object" is a generalization of the concept of "event", associated with a point in spacetime. An event, by definition is "instantonic" as well, because it occurs at one particular point in spacetime.

A configuration can be very complicated and self-referential, and thus being a record of the configurations at earlier values of s. In this respect this approach resembles that by Barbour [111], who considered "time capsules" with memory of the past. As a model, he considered a triangleland, whose configurations are triangles. Instead of triangleland, we consider here the Clifford space, in which configurations are modeled by oriented r-volumes ($r = 0, 1, 2, 3, 4$) in spacetime. In this respect our model differs from Barbour's model, in which the triangles are in 3-dimensional space. Instead of 3-dimensional space, I consider a 4-dimensional space with signature $(+ - - -)$. During the development of physics it was recognized that a 3-dimensional space is not suitable for formulation of the theory describing the physical phenomena, such as electromagnetism and moving objects. In other words, the theory of relativity requires 4-dimensional space, with an extra dimension x^0, whose signature is opposite to the signature of three spatial dimensions. The fourth dimension was identified with time, $x^0 \equiv t$. Such identification, though historically very useful, has turned out to be misleading [49, 112–122]. In fact, x^0 is not the true time, it is just a coordinate of the fourth dimension. The evolution time is something else. In the Stueckelberg theory [49, 112–122] its origin remains unexplained. In the approach with Clifford space, the evolution time (evolution parameter) is $s = (\sigma + \tilde{\sigma})/2$, i.e., a superposition of the *scalar* coordinate, σ, and the *pseudoscalar* coordinate, $\tilde{\sigma}$. This is the parameter that distinguishes configurations within a 1-dimensional family. In principle, the configurations can be very complicated and self-referential, including conscious experiences of an observer. Thus s distinguishes different conscious experiences of an observer [49,120]; it is the time experienced by a conscious observer. A wave function $\phi(x^{iM}) = e^{i\Lambda}\psi(s, x^M)$ "selects" in the vast space $\mathcal{C}$ of all possible configurations x^{iM} a subspace $\mathcal{S} \in \mathcal{C}$ of configurations. More precisely, ϕ assigns a probability density over the points of $\mathcal{C}$, so that some points are more likely to be experienced by an observer than the other points. In particular, $\phi(x^{iM})$ can be a localized wave packet evolving along s. For instance, such a wave packet can be localized around a worldline $x^{iM} = X_0^{iM}(s)$ in $\mathcal{C}$, which from the point of view of M_4, is a succession (evolution) of configurations X_0^M at different values of the parameter s. If configurations are complicated and include the external world and an observer's brain, such wave packet $\psi(s, x^M)$ determines the evolution of conscious experiences of an observer coupled by his sense organs to the external world.

5.3 Generators of Clifford algebras as quantum mechanical operators

5.3.1 *Orthogonal and symplectic Clifford algebras*

After having exposed a broader context of the role of Clifford algebras in physics, let me now turn to a specific case and consider the role of Clifford algebras in quantization. The inner product of generators of Clifford algebra gives the metric. We distinguish two cases:[2]

(i) If metric is *symmetric*, then the inner product is given by the *anticommutator* of generators; this is the case of an *orthogonal Clifford algebra*:

$$\frac{1}{2}\{\gamma_a, \gamma_b\} \equiv \gamma_a \cdot \gamma_b = g_{ab}. \tag{5.27}$$

(ii) If metric is *antisymmetric*, then the inner product is given by the commutator of generators; this is the case of a *symplectic Clifford algebra*:

$$\frac{1}{2}[q_a, q_b] \equiv q_a \wedge q_b = J_{ab}. \tag{5.28}$$

Here q_a are the symplectic basis vectors that span a *symplectic space*, whose points are associated with symplectic vectors [68]

$$z = z^a q_a. \tag{5.29}$$

---

[2]The topics of this section, apart from the end of Sec. 5.4.1 and Sec. 5.5, was first considered in Ref. [68].

Here z^a are commuting phase space coordinates,

$$z^a z^b - z^b z^a = 0. \tag{5.30}$$

An example of symplectic space in physics is *phase space*, whose points are coordinates and momenta of a particle:

$$z^a = (x^\mu, p^\mu) \equiv (x^\mu, \bar{x}^\mu) \equiv (x^\mu, x^{\bar{\mu}}). \tag{5.31}$$

The corresponding basis vectors then split according to

$$q_a = (q_\mu^{(x)}, q_\mu^{(p)}) \equiv (q_\mu, \bar{q}_\mu) \equiv (q_\mu, q_{\bar{\mu}}), \qquad \mu = 1, 2, ..., n, \tag{5.32}$$

and the relation (5.28) becomes

$$\tfrac{1}{2}[q_\mu^{(x)}, q_\nu^{(p)}] \equiv \frac{1}{2}[q_\mu, q_{\bar{\nu}}] = J_{\mu\bar{\nu}} = g_{\mu\nu},$$

$$[q_\mu^{(x)}, q_\nu^{(x)}] = 0, \qquad [q_\mu^{(p)}, q_\nu^{(p)}] = 0, \tag{5.33}$$

where we have set

$$J_{ab} = \begin{pmatrix} 0 & g_{\mu\nu} \\ -g_{\mu\nu} & 0 \end{pmatrix}. \tag{5.34}$$

Here, depending on the case considered, $g_{\mu\nu}$ is the euclidean, $g_{\mu\nu} = \delta_{\mu\nu}$, $\mu, \nu = 1, 2, ..., n$, or the Minkowski metric, $g_{\mu\nu} = \eta_{\mu\nu}$. In the latter case we have $\mu, \nu = 0, 1, 2, ..., n - 1$.

We see that (5.33) are just the Heisenberg commutation relations for coordinate and momentum operators, identified as[3]

$$\hat{x}_\mu = \frac{1}{\sqrt{2}} q_\mu^{(x)} \; ; \qquad \hat{p}_\mu = \frac{i}{\sqrt{2}} q_\mu^{(p)}. \tag{5.35}$$

Then we have

$$[\hat{x}_\mu, \hat{p}_\nu] = ig_{\mu\nu} \; , \qquad [\hat{x}_\mu, \hat{x}_\nu] = 0 \; , \qquad [\hat{p}_\mu, \hat{p}_\nu] = 0. \tag{5.36}$$

Instead of a symplectic vector $z = z^a q_a$, let us now consider another symplectic vector, namely

$$F = \frac{\partial f}{\partial z^a} q^a, \tag{5.37}$$

where $f = f(z)$ is a function of position in phase space. The wedge product of two such vectors is

$$F \wedge G = \frac{\partial f}{\partial z^a} q^a \wedge q^b \frac{\partial g}{\partial z^b} = \frac{\partial f}{\partial z^a} J^{ab} \frac{\partial g}{\partial z^b}, \tag{5.38}$$

where in the last step we used the analog of Eq. (5.28) for the reciprocal quantities $q^a = J^{ab} q_b$, where J^{ab} is the inverse of J_{ab}.

---

[3] We insert factor i in order to make the operator $\hat{p}_\mu$ hermitian.

Eq. (5.38) is equal to the Poisson bracket of two phase space functions. namely, using (5.31) and (5.34), we have

$$\frac{\partial f}{\partial z^a} J^{ab} \frac{\partial g}{\partial z^b} = \frac{\partial f}{\partial x^\mu} \eta^{\mu\nu} \frac{\partial g}{\partial p^\nu} - \frac{\partial f}{\partial p^\mu} \eta^{\mu\nu} \frac{\partial g}{\partial x^\nu} \equiv \{f, g\}_{PB}. \tag{5.39}$$

In particular, if

$$f = z^c, \qquad g = z^d, \tag{5.40}$$

Eqs. (5.38), (5.39) give

$$q^a \wedge q^b = J^{ab} = \{z^a, z^b\}_{PB}. \tag{5.41}$$

We see that the Heisenberg commutation relations for operators $\hat{x}^\mu$, $\hat{p}^\mu$ are obtained automatically, if we express the Poisson bracket relations in terms of the wedge product of the symplectic vectors

$$F = \frac{\partial f}{\partial z^a} q^a = \frac{\partial f}{\partial x^\mu} q^\mu_{(x)} + \frac{\partial f}{\partial p^\mu} q^\mu_{(p)} \quad \text{and} \quad G = \frac{\partial g}{\partial z^a} = \frac{\partial g}{\partial x^\mu} q^\mu_{(x)} + \frac{\partial f}{\partial p^\mu} q^\mu_{(p)}. \tag{5.42}$$

By having taken into account not only the coordinates and functions in a symplectic space, but also corresponding basis vectors, we have found that basis vectors are in fact quantum mechanical operators [68]. Moreover, the Poisson bracket between classical phase space variable, $\{z^a, z^b\}_{PB}$, is *equal* to the commutator, $\frac{1}{2}[q^a, q^b] = q^a \wedge q^b$, of vectors (i.e., of operators) q^a and q^b [68]. According to this picture, quantum operators are already present in the classical symplectic form, if we write the symplectic metric as the inner product of symplectic basis vectors. The latter vectors are just the quantum mechanical operators.

Analogous procedure can be performed with orthogonal Clifford algebras. Then a point in phase space can be described as a vector

$$\lambda = \lambda^a \gamma_a, \tag{5.43}$$

where λ^a are anticommuting phase space coordinates,

$$\lambda^a \lambda^b + \lambda^b \lambda^a = 0, \tag{5.44}$$

and γ_a basis vectors, satisfying Eq. (5.27). If we split the vectors γ_a and the metric γ_{ab} according to

$$\gamma_a = (\gamma_\mu, \bar{\gamma}_\mu), \qquad \mu = 0, 1, 2, ..., n-1, \tag{5.45}$$

$$g_{ab} = \begin{pmatrix} g_{\mu\nu} & 0 \\ 0 & g_{\mu\nu} \end{pmatrix} \tag{5.46}$$

and introduce a new basis, the so called *Witt basis*,

$$\theta_\mu = \frac{1}{\sqrt{2}}(\gamma^\mu + i\bar{\gamma}_\mu),$$

$$\bar{\theta}_\mu = \frac{1}{\sqrt{2}}(\gamma^\mu - i\bar{\gamma}_\mu), \tag{5.47}$$

then the Clifford algebra relations (5.27) become

$$\theta_\mu \cdot \bar{\theta}_\nu \equiv \frac{1}{2}\left(\theta_\mu\bar{\theta}_\nu + \bar{\theta}_\nu\theta_\mu\right) = \eta_{\mu\nu},$$

$$\theta_\mu \cdot \theta_\nu = 0, \qquad \bar{\theta}_\mu \cdot \bar{\theta}_\nu = 0. \tag{5.48}$$

These are the anticommutation relations for fermionic creation and annihilation operators.

Let us now introduce functions $\tilde{f}(\lambda)$ and $\tilde{g}(\lambda)$, and consider the vectors

$$\tilde{F} = \frac{\partial \tilde{f}}{\partial \lambda^a}\gamma^a, \qquad \tilde{g} = \frac{\partial \tilde{g}}{\partial \lambda^a}\gamma^a. \tag{5.49}$$

The dot product of those vectors is

$$\tilde{F} \cdot \tilde{G} = \frac{\partial \tilde{f}}{\partial \lambda^a}\gamma^a \cdot \frac{\partial \tilde{g}}{\partial \lambda^b}\gamma^b = \frac{\partial \tilde{f}}{\partial \lambda^a}g^{ab}\frac{\partial \tilde{g}}{\partial \lambda^b} = \{\tilde{f}, \tilde{g}\}_{PB}, \tag{5.50}$$

where $g^{ab} = \gamma^a \cdot \gamma^b$ is the inverse of g_{ab}.

Eq. (5.50) shows that the dot product, which in the orthogonal case corresponds to the inner product, is equal to the Poisson bracket of two phase space functions, now composed with the symmetric metric g^{ab}.

If

$$\tilde{f} = \lambda^c, \qquad \tilde{g} = \lambda^d, \tag{5.51}$$

Eq. (5.50) gives

$$\tilde{F} \cdot \tilde{G} = \gamma^c \cdot \gamma^d = g^{cd}, \tag{5.52}$$

which in the Witt basis read as the fermionic anticommutation relations (5.48). This means that the Poisson bracket between the (classical) phase space variables λ^a, λ^b is equal to the anticommutator of the "operators" γ^a and γ^b:

$$\{\lambda^a, \lambda^b\}_{PB} = \frac{1}{2}\{\gamma^a, \gamma^b\} = g^{ab}. \tag{5.53}$$

Again we have that the basis vectors behave as quantum mechanical operators.

5.3.2 Equations of motion for a particle's coordinates and the corresponding basis vectors

We will now consider [68] a point particle, described by the phase space action

$$I = \frac{1}{2} \int d\tau \, (\dot{x}^a J_{ab} z^b + z^a K_{ab} z^b), \tag{5.54}$$

where

$$\frac{1}{2} z^a K_{ab} z^b = H \tag{5.55}$$

is the Hamiltonian, the quantity K_{ab} being a symmetric $2n \times 2n$ matrix.

Variation of the action (5.54) with respect to z^a gives

$$\dot{z}^a = J^{ab} \frac{\partial H}{\partial z^b}, \tag{5.56}$$

which are the Hamilton equations of motion.

A solution of equation (5.56) is a trajectory z in phase space. We can consider a trajectory as an infinite dimensional vector with components $z^a(\tau) \equiv z^{a(\tau)}$. Here $a(\tau)$ is the index that denotes components; it is a double index, with a being a discrete index, and (τ) a continuous one. Corresponding basis vectors are $q_a(\tau) \equiv q_{a(\tau)}$, and they satisfy the relations

$$q_{a(\tau)} \wedge q_{b(\tau)} = J_{a(\tau)b(\tau)} = J_{ab}\delta(\tau - \tau'), \tag{5.57}$$

which are an extension of the relations (5.28) to our infinite dimensional case.

A trajectory is thus described by the vector

$$z = z^{a(\tau)} q_{a(\tau)} \equiv \int d\tau z^a(\tau) q_a(\tau). \tag{5.58}$$

The phase space velocity vector is

$$v = \dot{z}^{a(\tau)} q_{a(\tau)} = -z^{a(\tau)} \dot{q}_{a(\tau)}, \tag{5.59}$$

where we have assumed that the "surface" term vanishes:

$$v = \int d\tau \dot{z}^a(\tau) q_a(\tau) = - \int d\tau z^a(\tau) \dot{q}_a(\tau) + z^a(\tau) q_a(\tau) \Big|_{\tau_1}^{\tau_2}. \tag{5.60}$$

The last term vanishes if $z^a(\tau_2) q_a(\tau_2) = z^a(\tau_1) q_a(\tau_1)$.

The action (5.54) can be written as

$$I = \frac{1}{2} \left(\dot{z}^{a(\tau)} J_{a(\tau)b(\tau)} z^{b(\tau')} + z^{a(\tau)} K_{a(\tau)b(\tau')} z^{b(\tau')} \right), \tag{5.61}$$

where $J_{a(\tau)b(\tau)}$ is given in Eq. (5.57), and

$$K_{a(\tau)b(\tau')} = K_{ab}\delta(\tau - \tau'). \tag{5.62}$$

The corresponding equations of motion are

$$\dot{z}^{a(\tau)} = J^{a(\tau)c(\tau'')} K_{c(\tau'')b(\tau')} z^{b(\tau')}. \tag{5.63}$$

Multiplying both sides of the latter equation by $q_{a(\tau)}$, we obtain

$$\dot{z}^{a(\tau)} q_{a(\tau)} = -q^{a(\tau)} K_{a(\tau)b(\tau')} z^{b(\tau')}. \tag{5.64}$$

We have raised the index by $J^{a(\tau)c(\tau'')}$ and taken into account that $J^{a(\tau)c(\tau'')} = -J^{c(\tau'')a(\tau)}$. Eq. (5.64) is just Eq. (5.63), expressed in terms of the basis vectors. Both equations are equivalent.

Using the relation (5.59) in Eq. (5.64), we obtain

$$z^{b(\tau')} \dot{q}_{b(\tau')} = q^{a(\tau)} K_{a(\tau)b(\tau')} z^{b(\tau')}. \tag{5.65}$$

Apart from the surface term that we have neglected in Eq. (5.60), the last equation, (5.65), is equivalent to the classical equation of motion (5.56), only the τ-dependence has been switched from the components to the basis vectors.

A curious thing happens if we assume that Eq. (5.65) holds for an arbitrary trajectory (Fig. 5.6). Then, instead of (5.65), we can write

$$\dot{q}_{b(\tau')} = q^{a(\tau)} K_{a(\tau)b(\tau')}. \tag{5.66}$$

Inserting into the latter equation the explicit expression (5.62) for $K_{a(\tau)b(\tau')}$

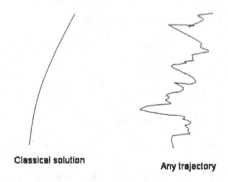

Classical solution

Any trajectory

Fig. 5.6 If the operator equations of motion (5.65) hold for any path $z^a(\tau)$ this means that coordinates and momenta are undetermined.

and writing $\dot{q}_{b(\tau)} = \dot{q}_b(\tau)$, $q^{a(\tau)} = q^a(\tau)$, we obtain

$$\dot{q}_a(\tau) = K_{ab} q^b(\tau). \tag{5.67}$$

This can be written as

$$\dot{q}_a = [q_a, \hat{H}], \tag{5.68}$$

where

$$\hat{H} = \frac{1}{2} q^a K_{ab} q^b, \tag{5.69}$$

is the Hamilton operator, satisfying

$$[q_a, \hat{H}] = K_{ab} q^b. \tag{5.70}$$

Starting from the classical action (5.54), we have arrived at the Heisenberg equations of motion (5.68) for the basis vectors q_a. On the way we have made a crucial assumption that the particle does not follow a trajectory $z^a(\tau)$ determined by the classical equations of motion, but that it can follow any trajectory. By the latter assumption we have passed from the classical to the quantized theory. We have thus found yet another way of performing quantization of a classical theory. Our assumption that a trajectory (a path) can be arbitrary, corresponds to that by Feynman path integrals. In our procedure we have shown how such an assumption of arbitrary path leads to the Heisenberg equations of motion for operators.

5.3.3 *Supersymmetrization of the action*

The action (5.54) can be generalized [68] so to contain not only a symplectic, but also an orthogonal part. For this purpose, we introduce the generalized vector space whose elements are

$$z = z^A q_A, \tag{5.71}$$

where

$$z^A = (z^a, \lambda^a), \qquad z^a = (x^\mu, \bar{x}^\mu), \qquad \lambda^a = (\lambda^\mu, \bar{\lambda}^\mu) \tag{5.72}$$
$$\text{\textit{symplectic part}} \qquad \text{\textit{orthogonal part}}$$

are coordinates, and

$$q_A = (q_a, \gamma_a), \qquad q_a = (q_\mu, \bar{q}_\mu), \qquad \gamma_a = (\gamma_\mu, \bar{\gamma}_\mu) \tag{5.73}$$
$$\text{\textit{symplectic part}} \qquad \text{\textit{orthogonal part}}$$

are basis vectors. The metric is

$$\langle q_A q_B \rangle_0 = G_{AB} = \begin{pmatrix} J_{ab} & 0 \\ 0 & g_{ab} \end{pmatrix}, \tag{5.74}$$

where $J_{ab} = -J_{ba}$ and $g_{ab} = g_{ba}$.

Let us consider a particle moving in such space. Its worldline is

$$z^A = Z^A(\tau). \tag{5.75}$$

An example of a possible action is

$$I = \frac{1}{2} \int d\tau \, \dot{Z}^A G_{AB} Z^B + \text{interaction terms.} \tag{5.76}$$

Using (5.72)–(5.74), the latter action can be split as

$$I = \frac{1}{2} \int d\tau \, (\dot{z}^a J_{ab} z^b + \dot{\lambda}^a g_{ab} \lambda) + \text{interaction terms}$$

$$= \frac{1}{2} \int d\tau \, (\dot{x}^\mu \eta_{\mu\nu} \bar{x}^\nu - \dot{\bar{x}}^\mu \eta_{\mu\nu} x^\nu + \dot{\lambda}^\mu \eta_{\mu\nu} \lambda^\nu + \dot{\bar{\lambda}} \eta_{\mu\nu} \bar{\lambda}^\nu)$$

$$+ \text{interaction terms} \tag{5.77}$$

Here z^a are commuting, and λ^a anticommuting (Grassmann) coordinates. The canonical momenta are

$$p_\mu^{(x)} = \frac{\partial L}{\partial \dot{x}^\mu} = \frac{1}{2} \eta_{\mu\nu} \dot{x}^\nu \,, \qquad p_\mu^{(\bar{x})} = \frac{\partial L}{\partial \dot{\bar{x}}^\mu} = -\frac{1}{2} \eta_{\mu\nu} x^\nu,$$

$$p_\mu^{(\lambda)} = \frac{\partial L}{\partial \dot{\lambda}^\mu} = \frac{1}{2} \eta_{\mu\nu} \lambda^\nu \,, \qquad p_\mu^{(\bar{\lambda})} = \frac{\partial L}{\partial \dot{\bar{\lambda}}^\mu} = \frac{1}{2} \eta_{\mu\nu} \bar{\lambda}^\nu. \tag{5.78}$$

Instead of the coordinates $\lambda^a = (\lambda^\mu, \bar{\lambda}^\mu)$, we can introduce the new coordinates

$$\lambda'^a = (\lambda'^\mu, \bar{\lambda}'^\mu) \,, \qquad \lambda'^\mu \equiv \xi^\mu = \frac{1}{\sqrt{2}} (\lambda^\mu - i\bar{\lambda}^\mu),$$

$$\bar{\lambda}'^\mu \equiv \bar{\xi}^\mu = \frac{1}{\sqrt{2}} (\lambda^\mu + i\bar{\lambda}^\mu), \tag{5.79}$$

in which the metric is

$$g'_{ab} = \gamma'_a \cdot \gamma'_b = \begin{pmatrix} 0 & \eta_{\mu\nu} \\ \eta_{\mu\nu} & 0 \end{pmatrix}. \tag{5.80}$$

In the new coordinates we have

$$\lambda^a g_{ab} \lambda^b = \lambda'^a g'_{ab} \lambda'^b = \dot{\xi}^\mu \eta_{\mu\nu} \bar{\xi}^\nu + \dot{\bar{\xi}}^\mu \eta_{\mu\nu} \xi^\nu. \tag{5.81}$$

Now the pairs of canonically conjugate variables are $(\xi^\mu, \frac{1}{2}\bar{\xi}_\mu)$ and $(\bar{\xi}^\mu, \frac{1}{2}\xi_\mu)$, whereas in the old coordinates the pairs were $(\lambda^\mu, \frac{1}{2}\lambda_\mu)$ and $(\bar{\lambda}^\mu, \frac{1}{2}\bar{\lambda}_\mu)$, which was somewhat unfortunate, because the variables in the pair were essentially the same.

The interaction term can be included by replacing the τ-derivative in the action (5.76) with the covariant derivative:

$$\dot{Z}^A \to \dot{Z}^A + A^A{}_C Z^C. \tag{5.82}$$

So we obtain [68]

$$I = \frac{1}{2} \int d\tau \, (\dot{Z}^A + A^A{}_C Z^C) G_{AB} Z^B. \tag{5.83}$$

This is a generalized Bars action [124], invariant under τ-dependent (local) rotations of Z^A. As discussed in [68], the gauge fields $A^A{}_C(\tau)$ are not dynamical; they have the role of Lagrange multipliers, whose choice determines a gauge, related to the way of how the canonically conjugated variables can be locally rotated into each other.

For a particular choice of $A^A{}_C$, we obtain

$$A^A{}_C Z^C G_{AB} Z^B = \alpha \, p^\mu p_\mu + \beta \, \lambda^\mu p_\mu + \gamma \, \bar{\lambda}^\mu p_\mu. \tag{5.84}$$

Here α, β, γ are Lagrange multipliers contained in $A^A{}_C$. Other choices of $A^A{}_C$ are possible, and they give expressions that are different from (5.84). A nice theory of how its works in the bosonic subspace, was elaborated by Bars (see, e.g., Refs. [124]).

The action (5.83), for the case (5.84), gives the constraints

$$p^\mu p_\mu = 0 \,, \qquad \lambda^\mu p_\mu = 0 \,, \qquad \bar{\lambda}^\mu p_\mu = 0, \tag{5.85}$$

or equivalently

$$p^\mu p_\mu = 0 \,, \qquad \xi^\mu p_\mu = 0 \,, \qquad \bar{\xi}^\mu p_\mu = 0, \tag{5.86}$$

if we use coordinates $\xi^a = (\xi^\mu, \bar{\xi})$, defined in Eq. (5.79).

Upon quantization we have

$$\hat{p}^\mu \hat{p}_\mu \Psi = 0 \,, \qquad \hat{\lambda}^\mu \hat{p}_\mu \Psi = 0 \,, \qquad \hat{\bar{\lambda}}^\mu \hat{p}_\mu \Psi = 0, \tag{5.87}$$

or equivalently

$$\hat{p}^\mu \hat{p}_\mu \Psi = 0 \,, \qquad \hat{\xi}^\mu \hat{p}_\mu \Psi = 0 \,, \qquad \hat{\bar{\xi}}^\mu \hat{p}_\mu \Psi = 0, \tag{5.88}$$

where the quantities with hat are are operators, satisfying

$$[\hat{x}^\mu, \hat{p}^\nu] = i\eta^{\mu\nu} \,, \qquad [\hat{x}^\mu, \hat{x}^\nu] = 0, \qquad [\hat{p}^\mu, \hat{p}^\nu] = 0, \tag{5.89}$$

$$\{\hat{\lambda}^\mu, \hat{\lambda}^\nu\} = 2i\eta^{\mu\nu} \,, \qquad \{\hat{\bar{\lambda}}^\mu, \hat{\bar{\lambda}}^\nu\} = 2i\eta^{\mu\nu} \,, \qquad \{\hat{\lambda}^\mu, \hat{\bar{\lambda}}^\nu\} = 0, \tag{5.90}$$

$$\{\hat{\xi}^\mu, \hat{\bar{\xi}}^\nu\} = \eta^{\mu\nu} \,, \qquad \{\hat{\xi}^\mu, \hat{\xi}^\nu\} = 0, \qquad \{\hat{\bar{\xi}}^\mu, \hat{\bar{\xi}}^\nu\} = 0, \tag{5.91}$$

The operators can be represented as

$$\hat{x}^\mu \to x^\mu \,, \qquad \hat{p}_\mu \to -i\frac{\partial}{\partial x^\mu} \,, \qquad \hat{\xi}^\mu \to \xi^\mu, \qquad \hat{\bar{\xi}}^\mu \to \frac{\partial}{\partial \xi^\mu}, \tag{5.92}$$

where

$$x^\mu x^\nu - x^\nu x^\mu = 0 \ , \qquad \xi^\mu \xi^\nu + \xi^\nu \xi^\mu = 0. \qquad (5.93)$$

A state Ψ can be represented as a wave function $\psi(x^\mu, \xi^\mu)$ of commuting coordinates x^μ and anticommuting (Grassmann) coordinates ξ^μ.

In Eq. (5.87) we have two copies of the Dirac equation, where $\hat{\lambda}^\mu$ and $\hat{\bar{\lambda}}^\mu$ satisfy the Clifford algebra anticommutation relations (5.90), and are related to γ^μ, $\hat{\gamma}^\mu$ according to

$$\hat{\lambda}^\mu = \gamma^\mu \ , \qquad \hat{\bar{\lambda}}^\mu = i \bar{\gamma}^\mu. \qquad (5.94)$$

Using (5.92), we find that the quantities γ_μ, $\bar{\gamma}_\mu$, satisfying

$$\gamma_\mu \cdot \gamma_\nu = \eta_{\mu\nu} \ , \qquad \bar{\gamma}_\mu \cdot \bar{\gamma}_\nu = \eta_{\mu\nu}. \qquad (5.95)$$

can be represented according to

$$\gamma_\mu = \frac{1}{\sqrt{2}} \left(\xi_\mu + \frac{\partial}{\partial \xi^\mu} \right) \ , \qquad \bar{\gamma}_\mu = \frac{1}{\sqrt{2}} \left(\xi_\mu - \frac{\partial}{\partial \xi^\mu} \right). \qquad (5.96)$$

If we expand $\psi(x^\mu, \xi^\mu)$ in terms of the Grassmann variables ξ^μ, we obtain a finite number (i.e., 2^n) of terms:

$$\psi(x^\mu, \xi^\mu) = \sum_{r=0}^{n} \psi_{\mu_1 \mu_2 \dots \mu_r} \, \xi^{\mu_1} \xi^{\mu_2} \dots \xi^{\mu_r}. \qquad (5.97)$$

In the case of 4D spacetime, $n = 4$, the wave function has $2^4 = 16$ components. The state Ψ can then be represented as a column $\psi^\alpha(x)$, $\alpha = 1, 2, .., 16$, and the operators γ^μ, $\bar{\gamma}^\mu$ as 16×16 matrices. Because we have built our theory over the $8D$ phase space, our spinor has not only four, but sixteen components. This gives a lot of room for unified theories of particles and fields [66, 70, 71, 125, 126].

5.4 Basis vectors, Clifford algebras, spinors and quantized fields

5.4.1 *Spinors as particular Clifford numbers*

We have seen that the generators of Clifford algebras have the properties of quantum mechanical operators. Depending on the kind of Clifford algebra, they satisfy the commutation or anti commutation relations for bosonic or fermionic creation and annihilation operators.

From the operators θ_μ and $\bar{\theta}_\mu$, defined in Eq. (5.45), we can build up spinors by taking a "vacuum"

$$\Omega = \prod_\mu \bar{\theta}_\mu \ , \qquad \text{which satisfies} \qquad \bar{\theta}_\mu \Omega = 0, \qquad (5.98)$$

and acting on it by "creation" operators θ_μ. So we obtain a "Fock space" basis for spinors [57, 59–61, 67]

$$s_\alpha = (1\Omega, \ \theta_\mu\Omega, \ \theta_\mu\theta_\nu\Omega, \ \theta_\mu\theta_\nu\theta_\rho\Omega, \ \theta_\mu\theta_\nu\theta_\rho\theta_\sigma\Omega), \tag{5.99}$$

in terms of which any state can be expanded as

$$\Psi_\Omega = \sum \psi^\alpha s_\alpha, \qquad \alpha = 1, 2, ..., 2^n. \tag{5.100}$$

Components ψ^α can be spacetime dependent fields. With the operators $\theta_\mu, \bar{\theta}_\mu$ we can construct spinors as the elements of a minimal left ideal of a Clifford algebra $Cl(2n)$. We will take the dimension of spacetime $n = 4$, so that our phase space will have dimension 8, and the Clifford algebra, built over it, will be $Cl(2, 6)$ which we will simply denote $Cl(8)$ or, in general, $Cl(2n)$.

Besides (5.98), there are other possible vacuums, e.g.,

$$\Omega = \prod_\mu \theta_\mu, \qquad \theta_\mu\Omega = 0, \tag{5.101}$$

$$\Omega = \left(\prod_{\mu \in R_1} \theta_\mu \right) \left(\prod_{\mu \in R_2} \bar{\theta}_\mu \right), \quad \theta_\mu\Omega = 0, \ \text{if } \mu \in R_2$$

$$\bar{\theta}_\mu\Omega = 0, \ \text{if } \mu \in R_2, \tag{5.102}$$

where

$$R_1 = \{\mu_1, \mu_2, ..., \mu_r\}, \qquad R_2 = \{\mu_{r+1}, \mu_{r+2}, ..., \mu_n\}. \tag{5.103}$$

There are 2^n vacuums of such a kind. By taking all those vacuums, we obtain the Fock space basis for the whole $Cl(2n)$. If $n = 4$, the latter algebra consists of 16 independent minimal left ideals, each belonging to a different vacuum (5.102) and containing 16-component spinors ($2^n = 16$ if $n = 4$), such as (5.100). A generic element of $Cl(8)$ is the sum of the spinors Ψ_{Ω_i}, $i = 1, 2, 3, 4, ..., 16$. belonging to the ideal associated with a vacuum Ψ_{Ω_i}:

$$\Psi = \sum_i \Psi_{\Omega_i} = \psi^{\alpha i} s_{\alpha i} \equiv \psi^{\tilde{A}} s_{\tilde{A}}, \qquad \tilde{A} = 1, 2, 3, 4, ..., 256, \tag{5.104}$$

where $s_{\tilde{A}} \equiv s_{\alpha i}$, $\alpha, i = 1, 2, ..., 16$, is the Fock space basis for $Cl(8)$, and $\psi^{\tilde{A}} \equiv \psi^{\alpha i}$ are spacetime dependent fields. The same element $\Psi \in Cl(8)$ can be as well expanded in terms of the multivector basis,

$$\Psi = \psi^A \gamma_A, \qquad A = 1, 2, 3, 4, ..., 256, \tag{5.105}$$

where

$$\gamma_A = \underline{1}, \ \gamma_{a_1}, \ \gamma_{a_1} \wedge \gamma_{a_2}, ..., \gamma_{a_1} \wedge \gamma_{a_2} \wedge ... \wedge \gamma_{a_{2n}}, \tag{5.106}$$

which can be written compactly as

$$\gamma_A = \gamma_{a_1} \wedge \gamma_{a_2} \wedge ... \wedge \gamma_{a_r} , \qquad r = 0, 1, 2, ..., 2n. \qquad (5.107)$$

We see that if we construct the Clifford algebra of the 8-dimensional phase space, then we have much more room for unification[4] of elementary particles and fields than in the case of $Cl(1,3)$, constructed over $4D$ spacetime. We have a state Ψ that can be represented by a 16×16 matrix, whose elements can represent all known particles of the 1st generation of the Standard Model. Thus, 64 elements of this 16×16 matrix include the left and right handed (L,R) versions of the states (e, ν_e), $(u, d)_r$, $(u, d)_b$, $(u, d)_g$, and their antiparticles, times factor two, because all those states, satisfying the generalized Dirac equation [70,71](see also Sec. 5.5.2), can in principle be superposed with complex amplitudes.

If we take *space inversion* (P) of those 64 states by using the same procedure as in Ref. [60], we obtain another 64 states of the 16×16 matrix representing Ψ, namely the states of mirror particles (P-particles). Under time reversal (T) (see Ref. [60]), we obtain yet another 64 states corresponding to time reversed particles (T-particles). And finally, under PT, we obtain 64 states of time reversed mirror particles (PT-particles). Altogether, we have $4 \times 64 = 256$ states:

$$s_{\alpha i} = \begin{pmatrix} \begin{pmatrix} e & \nu \\ \bar{e} & \bar{\nu} \end{pmatrix} & \begin{pmatrix} e & \nu \\ \bar{e} & \bar{\nu} \end{pmatrix}_P & \begin{pmatrix} e & \nu \\ \bar{e} & \bar{\nu} \end{pmatrix}_T & \begin{pmatrix} e & \nu \\ \bar{e} & \bar{\nu} \end{pmatrix}_{PT} \\ \begin{pmatrix} u & d \\ \bar{u} & \bar{d} \end{pmatrix}_r & \begin{pmatrix} u & d \\ \bar{u} & \bar{d} \end{pmatrix}_{r,P} & \begin{pmatrix} u & d \\ \bar{u} & \bar{d} \end{pmatrix}_{r,T} & \begin{pmatrix} u & d \\ \bar{u} & \bar{d} \end{pmatrix}_{r,PT} \\ \begin{pmatrix} u & d \\ \bar{u} & \bar{d} \end{pmatrix}_g & \begin{pmatrix} u & d \\ \bar{u} & \bar{d} \end{pmatrix}_{g,P} & \begin{pmatrix} u & d \\ \bar{u} & \bar{d} \end{pmatrix}_{g,T} & \begin{pmatrix} u & d \\ \bar{u} & \bar{d} \end{pmatrix}_{g,PT} \\ \begin{pmatrix} u & d \\ \bar{u} & \bar{d} \end{pmatrix}_b & \begin{pmatrix} u & d \\ \bar{u} & \bar{d} \end{pmatrix}_{b,P} & \begin{pmatrix} u & d \\ \bar{u} & \bar{d} \end{pmatrix}_{b,T} & \begin{pmatrix} u & d \\ \bar{u} & \bar{d} \end{pmatrix}_{b,PT} \end{pmatrix}, \qquad (5.108)$$

where

$$\begin{pmatrix} e & \nu \\ \bar{e} & \bar{\nu} \end{pmatrix} \equiv \begin{pmatrix} e_L & ie_L & \nu_L & i\nu_L \\ e_R & ie_R & \nu_R & i\nu_R \\ \bar{e}_L & i\bar{e}_L & \bar{\nu}_L & i\bar{\nu}_L \\ \bar{e}_R & i\bar{e}_R & \bar{\nu}_R & i\bar{\nu}_R \end{pmatrix}, \qquad (5.109)$$

and similarly for u, d.

---

[4]Unification based on Clifford algebras in phase space was considered by Zenczykowski [125].

Those states interact with the corresponding gauge fields[5], which include the gauge fields of the Standard Model, such as the photon, weak bosons and gluons. There exist also mirror versions, as well as T and PT versions of the standard gauge bosons[6].

Of the 256 particle states in Eq. (5.108), only 1/4 interact with our usual photons, whereas the remaining 3/4 do not interact with our photons, but they may interact with mirror photons, T-photons or PT-photons. This scheme thus predicts the existence of *dark matter*. If the matter in the universe were evenly distributed over the ordinary particles, P-particles, T-particles and PT-particles, then 1/4 of the matter would be visible, and 3/4 dark. In reality, the distribution of matter in the universe need not be even over the four different version of the particles. It can deviate from even distribution, but we expect that the deviation is not very big. According to the current astronomical observations about 81.7% of matter in the universe is dark, and only 18.3% is visible. This roughly corresponds to the ratio 1/4 of the "visible states" in matrix (5.108).

5.4.2 *Quantized fields as generalized Clifford numbers*

We can consider a field as an infinite dimensional vector. As an example, let us take

$$\Psi = \psi^{i(x)} h_{i(x)} \equiv \int \mathrm{d}^n x \, \psi^i(x) h_i(x), \qquad (5.110)$$

where $i = 1, 2$, $x \in \mathbb{R}^3$ or $x \in \mathbb{R}^{1,3}$ are, respectively, a discrete index, and (x) a continuous index, denoting, e.g., a point in 3D space, or an event in 4D spacetime. The infinite dimensional vector Ψ is decomposed with

---

[5] How this works in the case of $Cl(1,3)$ is shown in Ref. [60] (see also [70, 71]).

[6] Here we extend the concept of mirror particles and mirror gauge fields. The idea of mirror particles was first put forward by Lee and Yang [72] who realized that "...there could exist corresponding elementary particles exhibiting opposite asymmetry such that in the broader sense there will still be over-all right-left symmetry." Further they wrote: "If this is the case, it should be pointed out that there must exist two kinds of protons p_R and p_L, the right-handed one and the left-handed one." Lee and Yang thus considered the possibility of mirror particles, though they did not name them so, and as an example they considered ordinary and mirror protons. Later, Kobzarev et al. [73], instead of P-partners, considered CP-partners of ordinary particles and called them "mirror particles". They argued that a complete doubling of the known particles and forces, except gravity, was necessary. Subsequently, the idea of mirror particles has been pursued in Refs. [74–81]. The connection between mirror particles and dark matter was suggested by Blinnikov and Khlopov [76, 77], and later explored in many works, e.g., in [82–89]. For a review see [90]. An explanation of mirror particles in terms of algebraic spinors (elements of Clifford algebras) was exposed in Refs. [60, 67].

respect to an infinite dimensional basis, consisting of vectors $h_{i(x)} \equiv h_i(x)$, satisfying [68]

$$h_{i(x)} \cdot h_{j(x')} \equiv \frac{1}{2}(h_{i(x)}h_{j(x')} + h_{j(x')}h_{i(x)}) = \rho_{i(x)j(x')}, \qquad (5.111)$$

where $\rho_{i(x)j(x')}$ is the metric of the infinite dimensional space $\mathcal{S}$. The latter space may in general have non vanishing curvature [49]. If, in particular, the curvature of $\mathcal{S}$ is "flat", then we may consider a parametrization of $\mathcal{S}$ such that

$$\rho_{i(x)j(x')} = \delta_{ij}\delta(x - x'). \qquad (5.112)$$

In Eq. (5.111) we have a generalization of the Clifford algebra relations (5.4) to infinite dimensions.

Instead of the basis in which the basis vectors satisfy Eq. (5.111), we can introduce theWitt basis

$$h_{(x)} = \frac{1}{\sqrt{2}}(h_{1(x)} + ih_{2(x)}), \qquad (5.113)$$

$$\bar{h}_{(x)} = \frac{1}{\sqrt{2}}(h_{1(x)} - ih_{2(x)}), \qquad (5.114)$$

in which we have

$$h_{(x)} \cdot \bar{h}_{(x')} = \delta_{(x)(x')}, \qquad (5.115)$$

$$h_{(x)} \cdot h_{(x')} = 0 , \qquad \bar{h}_{(x)} \cdot \bar{h}_{(x')} = 0. \qquad (5.116)$$

The vector $h_{i(x)}$ and the corresponding components $\psi^{i(x)}$ may contain an implicit discrete index $\mu = 0, 1, 2, ..., n$, so that Eq. (5.110) explicitly reads

$$\Psi = \psi^{i\mu(x)}h_{i\mu(x)} = \psi^{\mu(x)}h_{\mu(x)} + \bar{\psi}^{\mu(x)}\bar{h}_{\mu(x)}. \qquad (5.117)$$

Then, Eqs. (5.115), (5.116) become the anticommuting relations for fermion fields:

$$h_{\mu(x)} \cdot \bar{h}_{\nu(x')} = \eta_{\mu\nu}\delta_{(x)(x')}, \qquad (5.118)$$

$$h_{\mu(x)} \cdot h_{\nu(x')} = 0 , \qquad \bar{h}_{\mu(x)} \cdot \bar{h}_{\nu(x')} = 0. \qquad (5.119)$$

The quantities $h_{\mu(x)}, \bar{h}_{\mu(x)}$ are a generalization to infinite dimensions of the Witt basis vectors θ_μ, $\bar{\theta}_\mu$, defined in Eq. (5.45).

Using $\bar{h}_{\mu(x)}$, we can define a vacuum state as the product [68]

$$\Omega = \prod_{\mu,x} \bar{h}_{\mu(x)} , \qquad \bar{h}_{\mu(x)}\Omega = 0. \qquad (5.120)$$

Then, using the definition (5.117) of a vector Ψ, we have

$$\Psi\Omega = \psi^{\mu(x)} h_{\mu(x)}\Omega. \tag{5.121}$$

Because $\bar{h}_{\mu(x)}\Omega = 0$, the second part of Ψ disappears in the above equation.

The infinite dimensional vector Ψ, defined in Eq. (5.117), consists of two parts, $\psi^{\mu(x)} h_{\mu(x)}$ and $\bar{\psi}^{\mu(x)} \bar{h}_{\mu(x)}$, which both together span the phase space of a field theory.

The vector $\psi^{\mu(x)} h_{\mu(x)}$ can be generalized to an element of an infinite dimensional Clifford algebra:

$$\psi_0 \underline{1} + \psi^{\mu(x)} h_{\mu(x)} + \psi^{\mu(x)\nu(x')} h_{\mu(x)} h_{\nu(x')} + ... \tag{5.122}$$

Acting with the latter object on the vacuum (5.120), we obtain

$$\Psi\Omega = (\psi_0 \underline{1} + \psi^{\mu(x)} h_{\mu(x)} + \psi^{\mu(x)\nu(x')} h_{\mu(x)} h_{\nu(x')} + ...)\Omega. \tag{5.123}$$

This state is the infinite dimensional space analog of the spinor as an element of a left ideal of a Clifford algebra. At a fixed point $x \equiv x^\mu$ there is no "sum" (i.e., integral) over x in expression (5.122), and we obtain a spinor with 2^n components. It is an element of a minimal left ideal of $Cl(2n)$. In 4D spacetime, $n = 4$, and we have $Cl(8)$ at fixed x.

Besides the vacuum (5.120) there are other vacuums, such as

$$\Omega = \prod_{\mu,x} h_{\mu(x)}, \qquad h_{\mu(x)}\Omega = 0, \tag{5.124}$$

and, in general,

$$\Omega = \left(\prod_{\mu \in R_1, x} \bar{h}_{\mu(x)}\right)\left(\prod_{\mu \in R_2, x} h_{\mu(x)}\right). \tag{5.125}$$

Here $R = R_1 \cup R_2$ is the set of indices $\mu = 0, 1, 2, ..., n$, and R_1, R_2 are subsets of indices, e.g., $R_1 = \{1, 3, 5, ..., n\}$, $R_2 = \{2, 4, ..., n-1\}$.

Expression (5.125) can be written as

$$\Omega = \prod_x \left(\prod_{\mu \in R_1} \bar{h}_{\mu(x)}\right)\left(\prod_{\mu \in R_2} h_{\mu(x)}\right) = \prod_x \Omega_{(x)}, \tag{5.126}$$

where

$$\Omega_{(x)} = \left(\prod_{\mu \in R_1} \bar{h}_{\mu(x)}\right)\left(\prod_{\mu \in R_2} h_{\mu(x)}\right), \tag{5.127}$$

is a vacuum at a fixed point x. At a fixed x, we have 2^n different vacuums, and thus 2^n different spinors, defined analogously to the spinor (5.123), belonging to different minimal ideal of $Cl(2n)$.

The vacuum (5.125) can be even further generalized by taking different domains $\mathcal{R}_1$, $\mathcal{R}_2$ of spacetime positions x:

$$\Omega = \left(\prod_{\mu \in \mathcal{R}_1, x \in \mathcal{R}_1} \bar{h}_{\mu(x)} \right) \left(\prod_{\mu \in \mathcal{R}_2, x \in \mathcal{R}_2} h_{\mu(x)} \right). \tag{5.128}$$

In such a way we obtain many other vacuums, depending on a partition of $\mathbb{R}^n$ into two domains $\mathcal{R}_1$ and $\mathcal{R}_2$ so that $\mathbb{R}^n = \mathcal{R}_1 \cup \mathcal{R}_2$.

Instead of the position space, we can take the momentum space, and consider, e.g., positive and negative momenta. In Minkowski spacetime we can have a vacuum of the form

$$\Omega = \left(\prod_{\mu, p^0 > 0, \mathbf{p}} \bar{h}_{\mu(p^0, \mathbf{p})} \right) \left(\prod_{\mu, p^0 < 0, \mathbf{p}} h_{\mu(p^0, \mathbf{p})} \right), \tag{5.129}$$

which is annihilated according to

$$\bar{h}_{\mu(p^0 > 0, \mathbf{p})} \Omega = 0, \qquad h_{\mu(p^0 < 0, \mathbf{p})} \Omega = 0. \tag{5.130}$$

For the vacuum (5.129), $\bar{h}_{\mu(p^0 > 0, \mathbf{p})}$ and $h_{\mu(p^0 < 0, \mathbf{p})}$ are *annihilation operators*, whereas $\bar{h}_{\mu(p^0 < 0, \mathbf{p})}$ and $h_{\mu(p^0 > 0, \mathbf{p})}$ are *creation operators* from which one can compose the states such as

$$\left(\psi_0 \underline{1} + \psi^{\mu(p^0 > 0, \mathbf{p})} h_{\mu(p^0 > 0, \mathbf{p})} + \psi^{\mu(p^0 > 0, \mathbf{p})} \psi^{\nu(p'^0 > 0, \mathbf{p}')} h_{\mu(p^0 > 0, \mathbf{p})} h_{\nu(p'^0 > 0, \mathbf{p}')} + \cdots \right.$$

$$\left. \cdots + \psi^{\mu(p^0 < 0, \mathbf{p})} \bar{h}_{\mu(p^0 < 0, \mathbf{p})} + \cdots \right) \Omega. \tag{5.131}$$

The vacuum, satisfying (5.130), has the property of the *bare* Dirac vacuum. This can be seen if one changes the notation according to

$$h_\mu(p^0 > 0, \mathbf{p}) \equiv b_\mu^\dagger(\mathbf{p}), \qquad \bar{h}_\mu(p^0 > 0, \mathbf{p}) \equiv b_\mu(\mathbf{p}), \tag{5.132}$$

$$\bar{h}_\mu(p^0 < 0, \mathbf{p}) \equiv d_\mu(\mathbf{p}), \qquad h_\mu(p^0 < 0, \mathbf{p}) \equiv d_\mu^\dagger(\mathbf{p}), \tag{5.133}$$

$$\Omega \equiv |0\rangle_{\text{bare}}. \tag{5.134}$$

A difference with the usual Dirac theory is that our operators have index μ which takes four values, and not only two values, but otherwise the principle is the same.

The operators $b_\mu^\dagger$ and b_μ, respectively, create and annihilate a positive energy fermion, whereas the operators d_μ, $d_\mu^\dagger$ create and annihilate a negative energy fermion. This is precisely a property of the bare Dirac vacuum. Instead of the bare vacuum, in quantum field theories we consider the the physical vacuum

$$|0\rangle = \prod_{\mu, \mathbf{p}} d_\mu(\mathbf{p}) |0\rangle_{\text{bare}}, \tag{5.135}$$

in which the negative energy states are filled, and which in our notation reads

$$\Omega_{\text{phys}} = \prod_{\mu, p^0 < 0, \mathbf{p}} \bar{h}_{\mu(p^0, \mathbf{p})} \Omega. \tag{5.136}$$

We see that in a field theory à la Clifford, a vacuum is defined as the product of fermionic operators (generators in the Witt basis). The Dirac (physical) vacuum is defined as a sea of negative energy states according to (5.135) or (5.136). Today it is often stated that the Dirac vacuum as the sea of negative energy states is an obsolete concept. But within a field theory based infinite dimensional Clifford algebras, a vacuum is in fact a "sea" of states defined by infinite (uncountable) product of operators.

With respect to the vacuum (5.129), one kind of particles are created by the positive energy operators $h_\mu(p^0 > 0, \mathbf{p})$, whilst the other kind of particles are created by the negative energy operator, $\bar{h}_\mu(p^0 < 0, \mathbf{p})$. The vacuum with reversed properties can also be defined, besides many other possible vacuums. All those vacuums participate in a description of the interactive processes of elementary particles. What we take into account in our current quantum field theory calculations seem to be only a part of a larger theory that has been neglected. It could be that some of the difficulties (e.g., infinities) that we have encountered in QFTs so far, are partly due to neglect of such a larger theory.

In an analogous way we can also construct [68] bosonic states as elements of an infinite dimensional symplectic Clifford algebra. The generators of the latter algebra are bosonic field operators. We will use them in the next subsection when constructing the action and field equations.

5.4.3 *The action and field equations*

A sympletic vector is [68]

$$\begin{aligned}
\Phi &= \phi^{i(x)} k_{i(x)} = \phi^{1(x)} k_{1(x)} + \phi^{2(x)} k_{2(x)} \\
&\equiv \phi^{(x)} k^{\phi}_{(x)} + \Pi_{(x)} k^{\Pi}_{(x)}, \\
& \qquad x \in \mathbb{R}^3 \quad \text{or} \quad x \in \mathbb{R}^{1,3}.
\end{aligned} \tag{5.137}$$

Here $\phi^{i(x)} = (\phi^{(x)}, \Pi^{(x)})$ are components and $k_{i(x)}$, $i = 1, 2$, basis vectors, satisfying

$$k_{i(x)} \wedge k_{j(x')} = \frac{1}{2} [k_{i(x)}, k_{j(x')}] = J_{i(x)j(x')}, \tag{5.138}$$

where

$$J_{i(x)j(x')} = \begin{pmatrix} 0 & \delta_{(x)(x')} \\ -\delta_{(x)(x')} & 0 \end{pmatrix}. \tag{5.139}$$

The action is

$$I = \int d\tau \left[\frac{1}{2} \dot{\phi}^{i(x)} J_{i(x)j(x')} \phi^{j(x')} - H \right], \tag{5.140}$$

where

$$H = \frac{1}{2} \phi^{i(x)} K_{i(x)j(x')} \phi^{j(x')} \tag{5.141}$$

is the Hamiltonian, and

$$\frac{1}{2} \dot{\phi}^{i(x)} J_{i(x)j(x')} \phi^{j(x')} = \frac{1}{2} (\Pi \dot{\phi} - \phi \dot{\Pi}) \tag{5.142}$$

the symplectic form.

In particular [68], if $x \equiv x^r \in \mathbb{R}^3$, $r = 1, 2, 3$, and

$$K_{i(x)j(x')} = \begin{pmatrix} (m^2 + \partial^r \partial_r)\delta(x - x') & 0 \\ 0 & \delta(x - x') \end{pmatrix}, \tag{5.143}$$

then we obtain the phase space action for a classical scalar field.

If $x \equiv x^r$, $r = 1, 2, 3$, and

$$K_{i(x)j(x')} = \left(-\frac{1}{2m} \partial^r \partial_r + V(x) \right) \delta(x - x') g_{ij}, \qquad g_{ij} = \begin{pmatrix} 0 & 1 \\ 1 & 0 \end{pmatrix}, \tag{5.144}$$

then the action (5.140) describes the classical Schrödinger field.

If $x \equiv x^\mu \in \mathbb{R}^{1,3}$, $\mu = 0, 1, 2, 3$, and

$$K_{i(x)j(x')} = \left(-\frac{1}{2\Lambda} \partial^\mu \partial_\mu \right) \delta(x - x') g_{ij}, \qquad g_{ij} = \begin{pmatrix} 0 & 1 \\ 1 & 0 \end{pmatrix}, \tag{5.145}$$

then from (5.140) we obtain the action for the classical Stueckelberg field.

From the action (5.140) we obtain the following equations of motion

$$\dot{\phi}^{i(x)} = J^{i(x)j(x')} \frac{\partial H}{\partial \phi^{j(x')}}, \tag{5.146}$$

where $\partial/\partial \phi^{j(x')} \equiv \delta/\delta \phi^{j(x')}$ is the functional derivative. By following the analogous procedure as in Sec. 5.3.2, we obtain, [68] the equations of motion for the operators:

$$\dot{k}_{j(x')} = k^{i(x)} K_{i(x)j(x')} = [k_{j(x')}, \hat{H}], \tag{5.147}$$

where

$$\hat{H} = \frac{1}{2} k^{i(x)} K_{i(x)j(x')} k^{j(x')}. \tag{5.148}$$

The Heisenberg equations of motion (5.147) can be derived from the action

$$I = \frac{1}{2} \int d\tau \left(\dot{k}^{i(x)} J_{i(x)j(x')} k^{j(x')} + k^{i(x)} K_{i(x)j(x')} k^{j(x')} \right). \tag{5.149}$$

The Poisson bracket between two functionals of the classical phase space fields is

$$\{f(\phi^{i(x)}), g(\phi^{j(x')})\}_{PB} = \frac{\partial f}{\partial \phi^{i(x)}} J^{i(x)j(x')} \frac{\partial g}{\partial \phi^{j(x')}}. \tag{5.150}$$

In particular, if $f = \phi^{k(x'')}$, $g = \phi^{\ell(x''')}$, Eq. (5.150) gives [68]

$$\{\phi^{k(x'')}, \phi^{\ell(x''')}\}_{PB} = J^{k(x'')\ell(x''')} = k^{k(x'')} \wedge k^{\ell(x''')} \equiv \frac{1}{2}[k^{k(x'')}, k^{\ell(x''')}]. \tag{5.151}$$

On the one hand, the Poisson bracket of two classical fields is equal to the symplectic metric. On the other hand, the symplectic metric is equal to the wedge product of basis vectors. In fact, the basis vectors are quantum mechanical operators, and they satisfy the quantum mechanical commutation relations

$$\frac{1}{2}[k_\phi(x), k_\Pi(x')] = \delta(x - x'), \tag{5.152}$$

or

$$[\hat{\phi}(x), \hat{\Pi}(x')] = i\delta(x - x'), \tag{5.153}$$

if we identify $\frac{1}{\sqrt{2}} k_\phi(x) \equiv \hat{\phi}(x)$, $\frac{i}{\sqrt{2}} k_\Pi(x') \equiv \hat{\Pi}(x')$

A similar procedure can be repeated for fermionic vectors [68].

5.5 Towards quantum gravity

5.5.1 *Gravitational field from Clifford algebra*

The generators of a Clifford algebra, γ_μ, $\bar{\gamma}_\mu$, are (i) tangent vectors to a manifold which, in particular, can be spacetime. On the other hand, (ii) the γ_μ, $\bar{\gamma}_\mu$ are superpositions of fermionic creation and annihilation operators, as shown in Eqs. (5.47), (5.48). This two facts, (i) and (ii), must have profound and far reaching consequences for quantum gravity. Here I am going to expose some further ingredients that in the future, after having been fully investigated, will illuminate the relation between quantum theory and gravity.

As a first step let us consider a generalized spinor field, defined in Sec. 5.4.1:

$$\Psi = \psi^{\tilde{A}} s_{\tilde{A}} = \phi^A \gamma_A. \tag{5.154}$$

We are interested in the expectation value of a a vector γ_μ with respect to the state Ψ:

$$\langle\gamma_\mu\rangle_1 \equiv \langle\Psi^\ddagger\gamma_\mu\Psi\rangle_1 = \langle\psi^{*\tilde{A}}s^\ddagger_{\tilde{A}}\gamma_\mu s_{\tilde{B}}\psi^{\tilde{B}}\rangle_1. \tag{5.155}$$

The subscript 1 means vector part of the expression. Recall from Sec. 5.2 that $\ddagger$ means reversion. Taking

$$\langle s^\ddagger_{\tilde{A}}\gamma_\mu s_{\tilde{B}}\rangle_1 = C^c_{\tilde{A}\tilde{B}\mu}\gamma_c , \tag{5.156}$$

we have

$$\langle\gamma_\mu\rangle_1 = e_\mu{}^c\gamma_c, \tag{5.157}$$

where

$$e_\mu{}^c = \psi^{*\tilde{A}}C^c_{\tilde{A}\tilde{B}\mu}\psi^{\tilde{B}} \tag{5.158}$$

is the vierbein. The vector γ_μ gives the flat spacetime metric

$$\gamma_\mu \cdot \gamma_\nu = \eta_{\mu\nu}. \tag{5.159}$$

The expectation value vector $\langle\gamma_\mu\rangle_1$ gives a curved spacetime metric

$$g_{\mu\nu} = \langle\gamma_\mu\rangle_1 \cdot \langle\gamma_\nu\rangle_1 = e_\mu{}^c e_\nu{}^d\eta_{cd} \tag{5.160}$$

which, in general, differs from $\eta_{\mu\nu}$. If $\psi^{\tilde{A}}$ depends on position $x \equiv x^\mu$ in spacetime, then also $e_\mu{}^c$ depends on x, and so does $g_{\mu\nu}$.

From Eq. (5.156) we obtain

$$e_\mu{}^a = \langle\gamma_\mu\rangle \cdot \gamma^a. \tag{5.161}$$

From the vierbein we can calculate the spin connection,

$$\omega_\mu{}^{ab} = \frac{1}{2}(e^{\rho b}e^a_{[\mu,\rho]} - e^{\rho a}e^b_{[\mu,\rho]} + e^{\rho b}e^{a\sigma}e_{\mu c}e^c_{[\sigma,\rho]}). \tag{5.162}$$

The curvature is

$$R_{\mu\nu}{}^{ab} = \partial_\mu\omega_\nu{}^{ab} - \partial_\nu\omega_\mu{}^{ab} + \omega_\mu{}^{ac}\omega_{\nu c}{}^b - \omega_\nu{}^{ac}\omega_{\mu c}{}^b. \tag{5.163}$$

In order to see whether the curvature vanishes or not, let us calculate $\omega^{ab}_{[\mu,\nu]}$ by using (5.158) in which we write

$$\psi^{*\tilde{A}}\psi^{\tilde{B}} \equiv \psi^{\tilde{A}\tilde{B}}. \tag{5.164}$$

We obtain

$$\begin{aligned}
\omega^{ab}_{[\mu,\nu]} = \frac{1}{2}\Big[&C^{b\rho}_{\tilde{A}\tilde{B}}C^a_{\tilde{C}\tilde{D}\mu}(\psi^{\tilde{A}\tilde{B}}\psi^{\tilde{C}\tilde{D}}{}_{,\rho})_{,\nu} - C^{b\rho}_{\tilde{A}\tilde{B}}C^a_{\tilde{C}\tilde{D}\rho}(\psi^{\tilde{A}\tilde{B}}\psi^{\tilde{C}\tilde{D}}{}_{,\mu})_{,\nu} \\
&-C^{a\rho}_{\tilde{A}\tilde{B}}C^b_{\tilde{C}\tilde{D}\mu}(\psi^{\tilde{A}\tilde{B}}\psi^{\tilde{C}\tilde{D}}{}_{,\rho})_{,\nu} + C^{a\rho}_{\tilde{A}\tilde{B}}C^b_{\tilde{C}\tilde{D}\rho}(\psi^{\tilde{A}\tilde{B}}\psi^{\tilde{C}\tilde{D}}{}_{,\mu})_{,\nu} \\
&+ \text{more terms} - (\mu \to \nu, \nu \to \mu)\Big].
\end{aligned} \tag{5.165}$$

The latter expression does not vanish identically. In general it could be different from zero, which would mean that also the curvature (5.163) is different from zero, and that the generalized spinor field $\psi^{\tilde{A}}(x)$ induces gravitation. This assertion should be checked by explicit calculations with explicit structure constants $C_{\tilde{A}\tilde{B}}^{\ b\rho}$ and/or their symmetry relations.

If $\psi^{\tilde{A}}(x)$ indeed induces gravitation, then we have essentially arrived at the basis of quantum gravity. At the basic level, gravity is thus caused by a spacetime dependent (generalized) spinor field $\psi^{\tilde{A}}(x)$ entering the expression (5.158) for vierbein. If $\psi^{\tilde{A}}(x)$ is constant, or proportional to $e^{ip_\mu x^\mu}$ which, roughly speaking, means that there is no non trivial matter, then $R_{\mu\nu}{}^{ab} = 0$. This has its counterpart in the (classical) Einstein's equation which say that matter curves spacetime.

5.5.2 *Fermion creation operators, branes as vacuums, branes with holes, and induced gravity*

The procedure described in Sec. 5.5.1 can be considered as a special case of quantized fields (5.123) at a fixed spacetime point x. We will now start from a generic object of the form (5.123). It consists of the terms such as

$$\psi^{\mu_1(x_1)\mu_2(x_2)...\mu_r(x_r)} h_{\mu_1(x_1)} h_{\mu_2(x_2)}...h_{\mu_r(x_r)}\Omega, \qquad (5.166)$$

where we assume that Ω is the vacuum given by Eq. (5.120). The operator $h_{\mu_i(x_i)}$ creates a fermion at a point x_i. The product of operators $h_{\mu_i(x_i)} h_{\mu_j(x_j)}$ creates a fermion at x_i and another fermion at x_j. By a generic expression (5.166) we can form any structure of fermions, e.g., a spin network. In the limit in which there are infinitely many densely packed fermions, we obtain arbitrary extended objects, such as strings, membranes, p-branes, or even more general objects, including instantonic branes, considered in Sec. 5.2.

Let us use the following compact notation for a state of many fermions forming an extended object in spacetime:

$$\left(\prod_{\mu, x \in \mathcal{R}} h_{\mu(x)} \right) \Omega. \qquad (5.167)$$

Here the product runs over spacetime points $x \in \mathcal{R}$ of a region $\mathcal{R}$ of spacetime M_D. In particular, $\mathcal{R}$ can be a p-brane's worldsheet V_{p+1}, whose parametric equation is $x^\mu = X^\mu(\sigma^a)$, $\mu = 0, 1, 2, ..., D-1$, $a = 1, 2, ..., p+1$, or it can be a brane-like instantonic object, also described by some functions

$X^\mu(\sigma)$. Then the product of operators in Eq. (5.167) can be written in the form

$$\prod_{\mu, x=X(\sigma)} h_{\mu(x)} \equiv h[X^\mu(\sigma)], \qquad (5.168)$$

where $h[X^\mu(\sigma)]$ is the operator that creates a brane or an instantonic brane (that we will also call "brane"). Here a brane is an extended objects consisting of infinitely many fermions, created according to

$$\psi_{\text{brane}} = h[X^\mu(\sigma)]\Omega = \left(\prod_{\mu, x=X(\sigma)} h_{\mu(x)} \right) \Omega. \qquad (5.169)$$

To make contact with the usual notation, we identify

$$\Omega \equiv |0\rangle \,, \qquad h[X^\mu(\sigma)] \equiv b^\dagger[X^\mu(\sigma)] \,, \qquad \psi_{\text{brane}} \equiv |X^\mu(\sigma)\rangle, \qquad (5.170)$$

and write

$$|X^\mu(\sigma)\rangle = b^\dagger[X^\mu(\sigma)]|0\rangle. \qquad (5.171)$$

A generic single brane state is a superposition of the brane states:

$$|\Psi\rangle = \int |X^\mu(\sigma)\rangle \mathcal{D}X(\sigma)\langle X^\mu(\sigma)|\Psi\rangle. \qquad (5.172)$$

In the notation of Eqs. (5.166)–(5.169), the latter expression reads

$$\Psi = \int \mathcal{D}X(\sigma)\psi[X^\mu(\sigma)]h[X^\mu(\sigma)]\Omega, \qquad (5.173)$$

where

$$\psi[X^\mu(\sigma)] = \lim_{r \to \infty, \Delta x_i \to 0} \psi^{\mu_1(x_1)...\mu_r(x_r)}. \qquad (5.174)$$

However, besides single brane states, there are also two-brane, three-brane, and in general, many-brane states. The brane Fock-space states are thus

$$b^\dagger[X_1^\mu(\sigma)]|0\rangle \,, \quad b^\dagger[X_1^\mu(\sigma)]b^\dagger[X_2^\mu(\sigma)]|0\rangle \,, \quad b^\dagger[X_1^\mu(\sigma)]...b^\dagger[X_r^\mu(\sigma)]|0\rangle \,, ...$$
$$(5.175)$$

A generic brane state is a superposition of those states.

If we act on the brane state (5.169) with the operator $\bar{h}_{\mu'(x')}$, we have

$$\bar{h}_{\mu'(x')}\Psi_{\text{brane}} = \bar{h}_{\mu'(x')} \left(\prod_{\mu, x=X(\sigma)} h_{\mu(x)} \right) \Omega. \qquad (5.176)$$

If x' is outside the brane, then nothing happens. But if x' is a position on the brane, then (5.176) is a state in which the particle at x' with the spin orientation μ is missing. In other words, (5.176) is a brane state with a hole at x'.

We may also form two hole state, many-hole states, and the states with a continuous set of holes,

$$\left(\prod_{\mu, x \in \mathcal{R}_1} \bar{h}_{\mu(x)} \right) \Psi_{\text{brane}} = \left(\prod_{\mu, x \in \mathcal{R}_1} \bar{h}_{\mu(x)} \right) \left(\prod_{\mu, x = X(\sigma)} h_{\mu(x)} \right) \Omega, \quad (5.177)$$

where $\mathcal{R}_1 \subset \mathcal{R} = \{X^\mu(\sigma)\}$. For instance, $\mathcal{R}_1$ can be a string or a brane of a lower dimensionality than the brane $X^\mu(\sigma^a)$.

If the space into which the brane is embedded has many dimensions, e.g., $D = 10 > p + 1$, then the brane's worldsheet V_{p+1} can represent our spacetime[7] which, if $p + 1 > 4$, has extra dimensions. The induced metric on V_{p+1} can be curved, and so we have curved spacetime. We have thus arrived at the brane world scenario. Holes in the brane are particles. More precisely, the point like holes in the worldsheet V_{p+1} are instantonic point particles, whereas the string like holes are instantonic strings, which can be either space like or time like (see Refs. [49, 127]).

Let me now outline how the induced metric on a brane V_{p+1} could be formally derived in terms of the operators $h_{\mu(x)}$, $\bar{h}_{\mu(x)}$. The corresponding operators in orthogonal basis are (see (5.113), (5.114)),

$$h_{1\mu(x)} = \frac{1}{\sqrt{2}} (h_{\mu(x)} + \bar{h}_{\mu(x)}), \quad (5.178)$$

$$h_{2\mu(x)} = \frac{1}{i\sqrt{2}} (h_{\mu(x)} - \bar{h}_{\mu(x)}), \quad (5.179)$$

satisfy the Clifford algebra relations

$$h_{i\mu(x)} \cdot h_{j\nu(x')} = \delta_{ij} \eta_{\mu\nu} \delta(x - x'). \quad (5.180)$$

In particular,

$$h_{1\mu(x)} \cdot h_{1\nu(x)} = \eta_{\mu\nu} \delta(0). \quad (5.181)$$

Comparing the latter result with

$$\gamma_\mu \cdot \gamma_\nu = \eta_{\mu\nu}, \quad (5.182)$$

we find that[8]

$$h_{1\mu(x)} = \gamma_\mu \sqrt{\delta(0)}. \quad (5.183)$$

---

[7]For more details on how an instantonic brane is related to our evolving spacetime, see Refs. [49, 127].

[8]Such notation could be set into a rigorous form if, e.g., in Eq. (5.180) we replace $\delta(x - x')$ with $\frac{1}{a\sqrt{\pi}} \exp[-\frac{(x-x')^2}{a^2}]$ and $\delta(0)$ with "$\delta(0)$" $\equiv \frac{1}{a\sqrt{\pi}}$. Then Eq. (5.181) is replaced by $h_{1\mu(x)} \cdot h_{1\nu(x)} = \eta_{\mu\nu}$ "$\delta(0)$". By inserting into the latter equation the relation $h_{1\mu(x)} = \gamma_\mu \sqrt{\text{"}\delta(0)\text{"}}$, we obtain $\gamma_\mu \cdot \gamma_\nu = \eta_{\mu\nu}$, which also holds in the limit $a \to 0$, because "$\delta(0)$" has disappeared from the equation.

This means that up to an infinite constant, $h_{1\mu(x)}$ is proportional to γ_μ, a basis vector of Minkowski spacetime. Thus, a proper renormalization of $h_{1\mu(x)}$ gives γ_μ.

In a given quantum state Ψ we can calculate the expectation value of $h_{i\mu(x)}$ according to

$$\langle h_{i\mu(x)} \rangle = \langle \Psi^\dagger h_{i\mu(x)} \Psi \rangle_1, \tag{5.184}$$

where the subscript 1 means vector part of the expression in the bracket. The inner product gives the expectation value of the metric:

$$\langle \rho_{i\mu(x)j\nu(x')} \rangle = \langle h_{i\mu(x)} \rangle \cdot \langle h_{j\nu(x')} \rangle. \tag{5.185}$$

This is the metric of an infinite dimensional manifold that, in general, is curved. In Refs. [49], a special case of such a manifold, for $i = j = 1$, called *membrane space* $\mathcal{M}$, was considered. It was shown how to define connection and curvature of $\mathcal{M}$.

Taking $i = j = 1$ and $x = x'$ in Eq. (5.185), we have

$$\langle \rho_{1\mu(x)\,1\nu(x)} \rangle = \langle h_{1\mu(x)} \rangle \cdot \langle h_{1\nu(x)} \rangle. \tag{5.186}$$

Upon renormalization according to (5.183) (see Footnote 8), we obtain

$$\langle g_{\mu\nu}(x) \rangle = \langle \gamma_\mu(x) \rangle \cdot \langle \gamma_\nu(x) \rangle, \tag{5.187}$$

where

$$\langle g_{\mu\nu}(x) \rangle = \langle \rho_{1\mu(x)1\nu(x)} \rangle \frac{1}{\sqrt{``\delta(0)"}} \tag{5.188}$$

is a position dependent metric of spacetime. We expect that the corresponding Riemann tensor is in general different from zero.

As an example let us consider the expectation value of a basis vector $h_{1\mu(x)}$ in the brane state (5.169):

$$\langle h_{1\mu(x)} \rangle = \langle \Psi^\dagger_{\text{brane}} h_{1\mu(x)} \Psi_{\text{brane}} \rangle_1 = \langle \Psi^\dagger_{\text{brane}} \frac{1}{\sqrt{2}} (h_{\mu(x)} + \bar{h}_{\mu(x)}) \Psi_{\text{brane}} \rangle_1. \tag{5.189}$$

From Eq. (5.176) in which the vacuum Ω is defined according to (5.120), we have

$$\bar{h}_{\mu(x)} \Psi_{\text{brane}} = \begin{cases} \Psi_{\text{brane}}(\breve{x}), & x \in \text{brane}; \\ 0, & x \notin \text{brane}. \end{cases} \tag{5.190}$$

Here $\Psi_{\text{brane}}(\breve{x})$, with the accent " $\breve{}$ " on x, denotes the brane with a hole at x. The notation $x \in$ brane means that x is on the brane, whereas $x \notin$ brane means that x is outside the brane created according to (5.169).

Because $(\bar{h}_{\mu(x)}\Psi_{\text{brane}})^{\ddagger} = \Psi_{\text{brane}}^{\ddagger}h_{\mu(x)}$, we also have

$$
\Psi_{\text{brane}}^{\ddagger}h_{\mu(x)} =
\begin{cases}
\Psi_{\text{brane}}(\check{x})^{\ddagger}, & x \in \text{brane}; \\
0, & x \notin \text{brane}.
\end{cases}
\tag{5.191}
$$

For the expectation value of $h_{1\mu(x)}$ we then obtain

$$
\langle h_{1\mu(x)} \rangle =
\begin{cases}
\frac{1}{\sqrt{2}}\langle \Psi_{\text{brane}}(\check{x})^{\ddagger}\Psi_{\text{brane}} \rangle_1 + \frac{1}{\sqrt{2}}\langle \Psi_{\text{brane}}^{\ddagger}\Psi_{\text{brane}}(\check{x}) \rangle_1, & x \in \text{brane}; \\
0, & x \notin \text{brane}.
\end{cases}
\tag{5.192}
$$

A similar expression we obtain for $\langle h_{2\mu(x)} \rangle$. The expectation value of the metric[9] (5.185) is

$$
\langle \rho_{i\mu(x)j\nu(x')} \rangle =
\begin{cases}
\langle \rho_{i\mu(x)j\nu(x')} \rangle |_{\text{brane}}, & \text{on the brane}; \\
0, & \text{outside the brane}.
\end{cases}
\tag{5.193}
$$

An interesting result is that outside the brane the expectation value of the metric is zero. Outside the brane, there is just the vacuum Ω. The expectation value of a vector $h_{i\mu(x)}$ in the vacuum, given by (5.120), is zero, and so is the expectation value $\langle \rho_{i\mu(x)j\nu(x')} \rangle$. This makes sense, because the vacuum Ω has no orientation that could be associated with a non vanishing effective vector. In Ω there also are no special points that could determine distances, and thus a metric. This is in agreement with the concept of configuration space, developed in Refs. [49], (see also Sec. 5.2), according to which outside a configuration there is no space and thus no metric: a physical space is associated with configurations, e.g., a system of particles, branes, etc.; without a configuration there is no physical space. In other words, a concept of a physical space unrelated to a configuration of physical objects has no meaning. Our intuitive believing that there exists a three (four) dimensional space(time) in which objects live is deceiving us. The three (four) dimensional space(time) is merely a subspace of the multidimensional configuration space of our universe, in which only position of a single particle is allowed to vary, while positions of all remaining objects are considered as fixed. Of course this is only an idealization. In reality, other objects are not fixed, and we have to take into account, when describing the universe, their configuration subspaces as well. Special and general relativity in 4-dimensional spacetime is thus a special case of a more

---

[9]Notice that the expectation value of the metric is *not* defined as $\langle \rho_{i\mu(x)j\nu(x')} \rangle = \langle \Psi^{\ddagger}\rho_{i\mu(x)j\nu(x')}\Psi \rangle$, but as $\langle \rho_{i\mu(x)j\nu(x')} \rangle = \langle h_{i\mu(x)} \rangle \cdot \langle h_{j\nu(x)} \rangle$.

general relativity in configuration space. Quantization of general relativity has failed, because it has not taken into account the concept of configuration space, and has not recognized that 4D spacetime is a subspace of the huge configuration space associated with our universe. The approach with quantized fields presented in this work has straightforwardly led us to the concept of many particle configurations and effective curved spaces associated with them.

If in Eq. (5.193) we take $i = j$, $x = x'$, and use Eqs. (5.186)–(5.188), then we obtain

$$\langle g_{\mu\nu}(x)\rangle = \begin{cases} g_{\mu\nu}(x)|_{\text{brane}} \neq 0, & \text{on the brane;} \\ 0, & \text{outside the brane.} \end{cases} \tag{5.194}$$

It is reasonable to expect that detailed calculations will give the result that $g_{\mu\nu}(x)|_{\text{brane}}$ is the induced metric on the brane, i.e.,

$$g_{\mu\nu}(x)|_{\text{brane}} = \partial_a X^\mu \partial_b X^\nu \eta_{\mu\nu} \equiv f_{ab}. \tag{5.195}$$

Recall that the brane can be our spacetime. We have thus pointed to a possible derivation of a curved spacetime metric from quantized fields in higher dimensions.

5.6   Quantized fields and Clifford space

In the previous section we considered fermion states that are generated by the action of creation operators on the vacuum Ω according to Eq. (5.166). In particular, a many fermion state can be a brane, formed according to Eq. (5.167). In Sec. 5.2 we showed that a brane can be approximately described by a polyvector (5.5) (see also (5.6)), which is a superposition of the Clifford algebra basis elements

$$1, \quad \gamma_\mu \wedge \gamma_\nu, \quad \gamma_\mu \wedge \gamma_\nu \wedge \gamma_\rho, \quad \gamma_\mu \wedge \gamma_\nu \wedge \gamma_\rho \wedge \gamma_\sigma. \tag{5.196}$$

This means that a Fock space element of the form (5.167) can be mapped into a polyvector:

$$\left(\prod_{\mu, x \in \mathcal{R}} h_{\mu(x)}\right) \Omega \longrightarrow x^M \gamma_M. \tag{5.197}$$

As an example let us consider the case in which the region $\mathcal{R}$ of spacetime is a closed line, i.e., a loop. The holographic projections of the area enclosed by the loop are given in terms of the bivector coordinates $X^{\mu\nu}$.

The loop itsel is described[10] by a bivectors $X^{\mu\nu}\gamma_\mu \wedge \gamma_\nu$. So we have the mapping

$$\left(\prod_{\mu,x\in\text{loop}} h_{\mu(x)} \right) \Omega \longrightarrow x^{\mu\nu}\gamma_{\mu\nu}. \qquad (5.198)$$

With the definite quantum states, described by Eq. (5.167) or (5.169) (see also (5.171)), which are the brane basis states, analogous to position states in the usual quantum mechanics, we can form a superposition (5.173) (see also (5.172)). To such an indefinite brane state there corresponds a state with indefinite polyvector coordinate X^M:

$$\int \mathcal{D}X(\sigma)\Psi[X(\sigma)]h[X(\sigma)]\Omega \longrightarrow \phi(x^M). \qquad (5.199)$$

In particular, if $h[X(\sigma)]\Omega$ is a loop, then we have the mapping

$$\int \mathcal{D}X(\sigma)\Psi[X(\sigma)]h[X(\sigma)]\Omega \longrightarrow \phi(x^{\mu\nu}). \qquad (5.200)$$

The circle is thus closed. With the mapping (5.197) we have again arrived at the polyvector $x^M\gamma_M$ introduced in Sec. 5.2. The polyvector coordinates x^M of a classical system satisfy the dynamics as formulated in Refs. [71, 104, 105]. That dynamics can be generalized to super phase space as discussed in Sec. 5.3, where besides the commuting coordinates x^μ, $\mu = 0, 1, 2, 3$, we introduced the Grassmann coordinates ξ^μ. In the quantized theory, the wave function $\psi(x^\mu, \xi^\mu)$ represents a 16-component field, ϕ^A, $A = 1, 2, ..., 16$, that depends on position x^μ in spacetime, and satisfies the Dirac equation

$$\gamma^\mu \partial_\mu \phi^A = 0, \qquad (5.201)$$

and the multicomponent Klein-Gordon equation

$$\eta^{\mu\nu} \partial_\mu \partial_\nu \phi^A = 0. \qquad (5.202)$$

In analogous way, besides commuting polyvector coordinates x^M, $M = 1, 2, ..., 16$, we have the corresponding Grassmann coordinates ξ^M, and the wave function $\phi(x^M)$ is generalized to $\phi(x^M, \xi^M)$. The expansion of $\phi(x^M, \xi^M)$ in terms of ξ^M gives a 2^{16}-component field, ϕ^A, $A = 1, 2, ..., 2^{16}$, that depends on position x^M in Clifford space, and satisfies the generalized Dirac equation

$$\gamma^M \partial_M \phi^A(x^M) = 0. \qquad (5.203)$$

---

[10] Of course, there is a class of loops, all having the same $X^{\mu\nu}$.

As the evolution parameter, i.e., the time along which the wave function evolves, we can take the time like coordinate x^0, or the time-like coordinate σ. Alternatively, we can take the light-like coordinate s, defined in Eq. (5.15), as the evolution parameter. Then, as shown in Ref. [109], the Cauchy problem can be well posed, in spite of the fact that in Clifford space there are eight time-like dimensions, besides eight space-like dimensions. Moreover, according to Refs. [45–47], there are no ghosts in such spaces, if the theory is properly quantized, and in Refs. [12, 25–29, 128] it was shown that the stability of solutions can be achieved even in the presence of interactions. We can now develop a theory of such quantized fields in Clifford space along similar lines as we did in Secs. 5.4 and 5.5 for the quantized fields in the ordinary spacetime. So we can consider the analog of Eqs. (5.167)–(5.195) and arrive at the induced metric on a 4-dimensional surface V_4 embedded in the 16-dimensional Clifford space. Whereas in Eqs. (5.167)–(5.195) we had ad hoc postulated the existence of extra dimensions, we now see that extra dimensions are incorporated in the configuration space of brane like objects created by the fermionic field operators $h_{\mu(x)}$. Our spacetime can thus be a curved surface embedded in such a configuration space.

5.7 Conclusion

Clifford algebras are very useful to describe extended objects as points in Clifford spaces, which are subspaces of configuration spaces. The Stueckelberg evolution parameter can be associated with the scalar and the pseudoscalar coordinate of the Clifford space.

The generators of orthogonal and symplectic Clifford algebras, i.e., the orthogonal and symplectic basis vectors, behave, respectively, as fermions and bosons. Quantization of a classical theory is the shift of description from components to the (orthogonal or symplectic) basis vectors.

We have found that a natural space to start from is a phase space, which can be either orthogonal or symplectic. We united both those phase spaces into a super phase space, whose points are described by anticommuting (Grassmann) and commuting coordinates, the basis vectors being the generators of orthogonal and symplecting Clifford algebras. We have considered the Clifford algebra $Cl(8)$ constructed over the 8-dimensional orthogonal part of the super phase space. Remarkably, the 256 spinor states of $Cl(8)$ can be associated with all the particles of the Standard Model, as well as with additional particles that do not interact with our photons and are therefore invisible to us. This model thus predicts dark matter. Moreover, it appears to be a promising step towards the unification of elementary particles and interactions (see also [70, 71, 105]).

Both, orthogonal and symplectic Clifford algebras can be generalized to infinite dimensions, in which case their generators (basis vectors) are bosonic and fermionic field creation and annihilation operators. In the Clifford algebra approach to field theories, a vacuum is the product of infinite, uncountable number of Fermionic field creation operators. They can form many sorts of possible vacuums as the seas composed of those field operators. In particular, strings and branes can be envisaged as being such seas. The field operators, acting on such brane states, can create holes in the branes, that behave as particles. From the expectation values of vector operators in such a one, two, or many holes brane state, we can calculate the metric on the brane. According to the brane world scenario, a brane can be our world. We have found that holes in a fermionic brane behave as particles, i.e., matter, in our world, and that the metric on the brane can be quantum mechanically induced by means of the fermionic creation and annihilation operators. We have thus found a road to quantum gravity that seems to avoid the usual obstacles. Moreover, this approach is along the lines, discussed in a seminal review article on quantum gravity by T. Padmanabhan [129], where strong theoretical arguments demonstrate that spacetime must have microscopic structure that at the macroscopic level exhibits thermal properties that can be described in thermodynamic langue. The microstructure considered in this chapter is composed of uncountably many fermionic field operators, considered as generators of an inifinite dimensional Clifford algebra.

Chapter 6

Brane Space and Branes as Conglomerates of Quantum Fields

In this chapter we will see how a Dirac-Nambu-Goto brane can be described as a point particle in an infinite dimensional space with a particular metric. This can be considered as a special case of a general theory in which branes are points in the brane space $\mathcal{M}$, whose metric is dynamical, just like in general relativity. Such a brane theory, amongst others, includes the flat brane space, whose metric is the infinite dimensional analog of the Minkowski space metric $\eta_{\mu\nu}$. A brane living in the latter space will be called "flat brane"; it is like a bunch of non interacting point particles. Quantization of the latter system leads to a system of non interacting quantum fields. Interactions can be included if we consider a non trivial metric in the space of fields. Then the effective classical brane is no longer a flat brane. For a particular choice of the metric in the field space we obtain the Dirac-Nambu-Goto brane. We also show how a Stueckelberg-like quantum field arises within the brane space formalism. With the Stueckelberg fields, we avoid certain well-known intricacies, especially those related to the position operator that is needed in our construction of effective classical branes from the systems of quantum fields[1].

6.1 Introduction

Relativistic membranes of arbitrary dimension (branes) [131–134], are very important objects in theoretical physics. An attractive possibility is a braneworld scenario [135–155] in which our spacetime is a 4-dimensional surface embedded in a higher dimensions space. Quantization of gravity

---

[1]This chapter is adapted version of the article [130] published by the author.

could then be achieved by quantizing the brane. Unfortunately, quantization of the Dirac-Nambu-Goto brane, satisfying the minimal surface action principle, is a tough problem that has not yet been solved in general. Although the quantization of the string, an extended object whose worldsheet has two dimensions, is rather well understood [156–158], this is not so in the case of branes with higher dimensional worldsheets (also called "world-volumes").

We will show how to solve this problem by considering the brane as a point in an infinite-dimensional brane space $\mathcal{M}$ that in general can be curved. The idea is that the metric of $\mathcal{M}$ is dynamical, just like in general relativity [49, 159, 160]. In particular the metric of $\mathcal{M}$ can be such that it gives the *Dirac-Nambu-Goto brane*, which is just the usual "minimal surface" brane. For other choices of $\mathcal{M}$-space metric we have branes that differ from the Dirac-Nambu-Goto brane, i.e., they do not satisfy the minimal surface action principle, but some other action principle. In particular, the $\mathcal{M}$-space metric can be "flat", which means that at any point of $\mathcal{M}$ it can be cast into the diagonal form. Then we have a brane analogue of a point particle in flat spacetime. Such a brane, from now on called *flat brane*, sweeps a worldsheet that is a bunch of straight worldlines (Fig. 6.1).

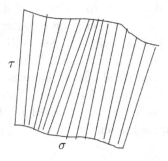

Fig. 6.1   A schematic illustration of a "flat brane". For some exact plots see Fig. 6.2.

A flat brane is thus just like continuous system of point particles. Quantization of a flat brane then leads to a system of non interacting quantum fields, $\varphi^r(x)$. The index r distinguishes one field from another, and because in the classical theory we had a continuous set of point particles, r must be continuous. If we consider interactions among the quantum fields, $\lambda_{rs}\varphi^r(x)\varphi^s(x)$, then the effective classical theory gives a brane living in a curved brane space, which, in particular, can be the Dirac-Nambu-Goto brane [160].

Instead of one brane a system can consist of many branes. Within the framework of such an enlarged configuration space it is possible to formulate the Stueckelberg quantum field theory with an invariant evolution parameter. We show how the latter parameter is embedded in the system's configuration.

Each of those branes can be described with a finite number of degrees of freedom, namely, with the center of mass, and additional degrees of freedom that take into account finite extension of the brane. Such extra degrees of freedom can be the coordinates of the Clifford space [99–101,106] that includes scalars, oriented lengths, areas, volumes and 4-volumes (pseudoscalars). In describing a multi brane system we can choose one brane and sample it with the coordinates of Clifford space, while for the remaining branes we retain the description with embedding functions. Clifford space is a 16-dimensional ultrahyperbolic space with neutral signature (8,8). From the scalar and pseudoscalar coordinates that span a 2-dimensional subspace with signature $(+-)$, we can form, with a suitable superposition, the analog of the light cone coordinates. In ultrahyperbolic spaces the Cauchy problem in general cannot be well posed, unless we determine initial data on the "light cone". In such a way we obtain the generalized Stueckelberg description [49, 112–122] of particles and branes, both classical and quantum. The Stueckelberg theory is based on the introduction of an evolution parameter τ, which is invariant under Lorentz transformations. In the literature we can find various explanations about the physical origin of τ, but none is generally accepted. In the approach pursued in this and in a series of previous papers [49, 122], the evolution parameter τ is a superposition of the scalar and pseudoscalar coordinate of the Clifford space, and is thus embedded in the configuration of the chosen brane. The latter brane, which in fact need not be just a brane, but whatever extended object that can be sampled as a brane, thus serves as a clock with which we measure the motion of the remaining branes that form the considered system.

6.2   Brane as a point in the brane space $\mathcal{M}$

The Dirac-Nambu-Goto brane is described by the minimal surface action

$$I = \kappa \int \mathrm{d}^{p+1}\xi \, (-\gamma)^{1/2}, \qquad (6.1)$$

where $\gamma \equiv \det \gamma_{ab}$, $\gamma_{ab} \equiv \partial_a X^\mu \partial_b X_\mu$ is the determinant of the worldsheet embedding functions $X^\mu(\xi^a)$, $\mu = 0, 1, 2, ..., D - 1$, $a = 0, 1, 2, ..., p$.

An equivalent action is the Schild action [161]

$$I_{\text{Schild}} = \frac{\kappa}{2k} \int d^{p+1}\xi \, (-\gamma), \tag{6.2}$$

from which as a consequence of equations of motion we obtain

$$\partial_a(-\gamma) = 0. \tag{6.3}$$

The determinant of the induced metric γ is thus a constant whose choice determines a gauge. We will choose a gauge such that

$$-\gamma = k^2, \tag{6.4}$$

in which case the canonical momentum $\pi_\mu{}^c = \kappa\sqrt{-\gamma}\,\partial^c X_\mu$ derived from the action (6.1) coincides with the canonical momentum $\kappa(-\gamma)\partial^c X_\mu/k$ derived from the action (6.2). Additionally, we can choose a gauge so that the determinant factorizes according to

$$(-\gamma) = \dot{X}^2(-\bar{\gamma}), \tag{6.5}$$

where $\dot{X}^2 \equiv \dot{X}^\mu \dot{X}_\mu$, $\dot{X}^\mu \equiv \partial X^\mu/\partial\tau$, and $\bar{\gamma} \equiv \det \partial_{\bar{a}} X^\mu \partial_{\bar{b}} X_\mu$, $\bar{a}, \bar{b} = 1, 2, ..., p$, the worldsheet parameters being split as $\xi^a = (\tau, \xi^{\bar{a}})$. Then instead of (6.2) we have

$$I_{\text{Schild}} = \frac{\kappa}{2k} \int d\tau \, d^p\sigma \, \dot{X}^2(-\bar{\gamma}), \tag{6.6}$$

which can be written as

$$I_{\text{Schild}} = \frac{\kappa}{2k} \int d\tau \, d^p\sigma \, d^p\sigma' \, (-\bar{\gamma}) \, \rho_{\mu\nu}(\sigma, \sigma') \, \dot{X}^\mu(\tau, \sigma) \dot{X}^\nu(\tau, \sigma'), \tag{6.7}$$

which is the τ-integral of a quadratic form in an infinite dimensional space $\mathcal{M}$ with the metric

$$\rho_{\mu\nu}(\sigma, \sigma') = (-\bar{\gamma}) \, \eta_{\mu\nu} \, \delta(\sigma - \sigma'). \tag{6.8}$$

At this point it is convenient to introduce a compact notation

$$\dot{X}^{\mu(\sigma)}(\tau) \equiv \dot{X}^\mu(\tau, \sigma), \qquad \rho_{\mu(\sigma)\nu(\sigma)} \equiv \rho_{\mu\nu}(\sigma, \sigma'), \tag{6.9}$$

and write [160]

$$I_{\text{Schild}} = \frac{\kappa}{2k} \int d\tau \, \rho_{\mu(\sigma)\nu(\sigma')} \dot{X}^{\mu(\sigma)}(\tau) \dot{X}^{\nu(\sigma')}(\tau). \tag{6.10}$$

Here we use the generalization of Einstein's summation convention, so that not only summation over the repeated indices μ, ν, but also the integration over the repeated continuous indices (σ), (σ') is assumed. Indices are lowered and raised, respectively, by $\rho_{\mu(\sigma)\nu(\sigma')}$ and its inverse $\rho^{\mu(\sigma)\nu(\sigma')}$. The infinite dimensional space $\mathcal{M}$ is called *brane space*, because its points $x^{\mu(\sigma)}$ represent kinematically possible branes [49, 159].

The quadratic form $\rho_{\mu(\sigma)\nu(\sigma')}\dot{X}^{\mu(\sigma)}(\tau)\dot{X}^{\nu(\sigma')}(\tau)$ is invariant under diffeomorphisms in the brane space $\mathcal{M}$. A curve in $\mathcal{M}$ is given by the parametric equation

$$x^{\mu(\sigma)} = X^{\mu(\sigma)}(\tau), \tag{6.11}$$

where $X^{\mu(\sigma)}(\tau)$ are τ-dependent functions. The velocity of a "point particle" in $\mathcal{M}$ is $\dot{X}^{\mu(\sigma)} \equiv \frac{\partial X^{\mu(\sigma)}}{\partial \tau}$.

The canonical momentum belonging to the action (6.10) is

$$p_{\mu(\sigma)} = \frac{\kappa}{k}\,\rho_{\mu(\sigma)\nu(\sigma')}\dot{X}^{\nu(\sigma')} = \frac{\kappa}{k}\,\dot{X}_{\mu(\sigma)} = \frac{\kappa}{k}\,(-\bar{\gamma})\dot{X}_{\mu}(\sigma). \tag{6.12}$$

Its contravariant components are

$$p^{\mu(\sigma)} = \rho^{\mu(\sigma)\nu(\sigma')}p_{\nu(\sigma')} = \frac{p^{\mu}(\sigma)}{(-\bar{\gamma})}, \tag{6.13}$$

where $p^{\mu}(\sigma) = \eta^{\mu\nu}p_{\nu}(\sigma)$.

The canonical momentum associated with the action (6.6) is

$$p_{\mu}(\sigma) = \frac{\kappa(-\bar{\gamma})\dot{X}_{\mu}}{k} = \frac{\kappa\sqrt{-\bar{\gamma}}\dot{X}_{\mu}}{\sqrt{\dot{X}^{\mu}\dot{X}_{\mu}}}, \tag{6.14}$$

where we have taken into account

$$k = \sqrt{\dot{X}^2}\sqrt{-\bar{\gamma}}, \tag{6.15}$$

which follows from (6.4) and (6.5). Using the latter equation (6.15), we verify that both momenta, (6.12) and (6.14), are equal, as they should be. In our notation $p_{\mu(\sigma)} = p_{\mu}(\sigma)$, whereas $p^{\mu(\sigma)}$ is given by Eq. (6.13).

The momentum $p_{\mu}(\sigma)$ satisfies the following constraint:

$$p_{\mu}(\sigma)p^{\mu}(\sigma) = \eta^{\mu\nu}p_{\mu}(\sigma)p_{\nu}(\sigma) = \kappa^2(-\bar{\gamma}). \tag{6.16}$$

We can also form the quadratic form of the momenta in $\mathcal{M}$-space,

$$p_{\mu(\sigma)}p^{\mu(\sigma)} = \rho^{\mu(\sigma)\nu(\sigma')}p_{\mu(\sigma)}p_{\nu(\sigma')} = \tilde{\kappa}^2, \tag{6.17}$$

in which the integration over repeated indices (σ) and (σ') is assumed. Comparing (6.17) and (6.16), we obtain

$$\tilde{\kappa}^2 = \int \kappa^2 \, \mathrm{d}\sigma. \tag{6.18}$$

Let us introduce the quantity

$$\tilde{k}^2 = \rho_{\mu(\sigma)\nu(\sigma')}\dot{X}^{\mu(\sigma)}\dot{X}^{\nu(\sigma')} = \int \mathrm{d}\sigma(-\bar{\gamma})\dot{X}^{\mu}\dot{X}^{\nu} = \int k^2 \mathrm{d}\sigma, \tag{6.19}$$

and take into account that Eqs. (6.18) and (6.19) imply $\tilde{\kappa}/\tilde{k} = \kappa/k$, i.e.,

$$\frac{\tilde{\kappa}}{\sqrt{\dot{X}^{\mu(\sigma)}\dot{X}_{\mu(\sigma)}}} = \frac{\kappa}{\sqrt{(-\bar{\gamma})\dot{X}^{\mu}\dot{X}_{\mu}}}. \tag{6.20}$$

Then we can write the Schild action (6.10) in terms of the quantities $\tilde{\kappa}$ and $\tilde{k}$:

$$I_{\text{Schild}} = \frac{\tilde{\kappa}}{2\tilde{k}} \int d\tau \, \rho_{\mu(\sigma)\nu(\sigma')}\dot{X}^{\mu(\sigma)}(\tau)\dot{X}^{\nu(\sigma')}(\tau). \tag{6.21}$$

The latter action is just a gauge fixed action obtained from the action [49, 159, 160]

$$I = \tilde{\kappa} \int d\tau \left(\rho_{\mu(\sigma)\nu(\sigma)}\dot{X}^{\mu(\sigma)}(\tau)\dot{X}^{\nu(\sigma')}(\tau) \right)^{1/2}, \tag{6.22}$$

which gives a minimal length worldline, i.e., a geodesic in $\mathcal{M}$-space. Indeed, the equations of motion derived from the action are [160]

$$\frac{\partial I}{\partial X^{\mu(\sigma)}} = \tilde{\kappa} \frac{d}{d\tau} \left(\frac{\dot{X}_{\mu(\sigma)}}{\sqrt{\dot{X}^2}} \right) - \frac{\tilde{\kappa}}{2} \partial_{\mu(\sigma)}\rho_{\alpha(\sigma')\beta(\sigma'')} \frac{\dot{X}^{\alpha(\sigma')}\dot{X}^{\beta(\sigma'')}}{\sqrt{\dot{X}^2}} = 0, \tag{6.23}$$

where

$$\dot{X}^2 \equiv \dot{X}^{\mu(\sigma)}\dot{X}_{\mu(\sigma)} = \rho_{\mu(\sigma)\nu(\sigma')}\dot{X}^{\mu(\sigma)}\dot{X}^{\nu(\sigma')}. \tag{6.24}$$

We use the following notation for functional derivatives:

$$\partial_{\mu(\sigma)} \equiv {}_{,\mu(\sigma)} \equiv \frac{\partial}{\partial x^{\mu(\sigma)}} \equiv \frac{\delta}{\delta x^{\mu}(\sigma)}. \tag{6.25}$$

Using $\dot{X}_{\mu(\sigma)} = \rho_{\mu(\sigma)\nu(\sigma)}\dot{X}^{\nu(\sigma)}$, and introducing the connection in $\mathcal{M}$,

$$\Gamma^{\mu(\sigma)}_{\alpha(\sigma')\beta(\sigma'')} = \frac{1}{2}\rho^{\mu(\sigma)\gamma(\sigma''')}(\rho_{\gamma(\sigma''')\alpha(\sigma'),\beta(\sigma'')} + \rho_{\gamma(\sigma''')\beta(\sigma''),\alpha(\sigma')}$$

$$-\rho_{\alpha(\sigma')\beta(\sigma''),\gamma(\sigma''')}), \tag{6.26}$$

we can write Eq. (6.23) in the form

$$\frac{1}{\sqrt{\dot{X}^2}}\frac{d}{d\tau}\left(\frac{\dot{X}^{\mu(\sigma)}}{\sqrt{\dot{X}^2}}\right) + \Gamma^{\mu(\sigma)}_{\alpha(\sigma')\beta(\sigma'')}\frac{\dot{X}^{\alpha(\sigma')}\dot{X}^{\beta(\sigma'')}}{\dot{X}^2} = 0, \tag{6.27}$$

which is the equation of geodesic in $\mathcal{M}$. If we insert into the latter equation the metric (6.8), then we obtain the equation of motion for the Dirac-Nambu-Goto brane. Equivalently, if we insert the metric (6.8) into the action (6.22) we obtain

$$I = \tilde{\kappa} \int d\tau \, \mathcal{L}[\dot{X}^{\mu}(\sigma), X^{\mu}(\sigma)], \tag{6.28}$$

where the Lagrangian

$$\mathcal{L}[\dot{X}^\mu(\sigma), X^\mu(\sigma)] = \left(\int \mathrm{d}^p \sigma \, (-\bar{\gamma}) \dot{X}^2 \right)^{1/2} \tag{6.29}$$

is a functional of infinite dimensional velocities and coordinates. From the Euler-Lagrange equations

$$\frac{\mathrm{d}}{\mathrm{d}\tau} \frac{\delta \mathcal{L}}{\delta \dot{X}^\mu(\sigma)} - \frac{\delta \mathcal{L}}{\delta X^\mu(\sigma)} = 0 \tag{6.30}$$

we obtain

$$\frac{\mathrm{d}}{\mathrm{d}\tau} \left(\frac{\tilde{\kappa}}{\sqrt{\dot{X}^2}} (-\bar{\gamma}) \dot{X}_\mu \right) + \partial_{\bar{a}} \left(\frac{\tilde{\kappa}(-\bar{\gamma}) \dot{X}^2 \partial^{\bar{a}} X_\mu}{\sqrt{\dot{X}^2}} \right) = 0, \tag{6.31}$$

where $\dot{X}^2 \equiv \dot{X}^{\mu(\sigma)} \dot{X}_{\mu(\sigma)} = \int \mathrm{d}^p \sigma \, (-\bar{\gamma}) \dot{X}^2$. Inserting Eq. (6.20) into the latter equation, we obtain

$$\frac{\mathrm{d}}{\mathrm{d}\tau} \left(\frac{\kappa \sqrt{-\bar{\gamma}}}{\sqrt{\dot{X}^2}} \dot{X}_\mu \right) + \partial_{\bar{a}} \left(\kappa \sqrt{-\bar{\gamma}} \sqrt{\dot{X}^2} \partial^{\bar{a}} X_\mu \right) = 0. \tag{6.32}$$

The same equation follows from the Dirac-Nambu-Goto action (6.1) in a gauge (6.5).

   We have arrived at the minimal length action (6.22) by using a particular metric (6.8). However, once we have such an action, we can assume that the metric need not be of that particular form. We can generalize the validity of the action (6.22) and the corresponding geodesic equation to any metric. In fact, we can assume that the metric of $\mathcal{M}$ is dynamical, like in general relativity, and that to the action (6.22) we have to add a kinetic term for the metric $\rho_{\mu(\sigma)\nu(\sigma')}$. An approach along such lines was investigated in Ref. [49]. Within such a generalized theory the metric (6.8), leading to the usual Dirac-Nambu-Goto brane, is just one of many other possible metrics, including the metric that is the brane space analog of the flat spacetime metric $\eta_{\mu\nu}$.

6.3   Special case: Flat brane space $\mathcal{M}$

The brane theory simplifies significantly if into the action (6.22) we plug the metric

$$\rho_{\mu(\sigma)\nu(\sigma')} = \eta_{\mu(\sigma)\nu(\sigma')} = \eta_{\mu\nu} \delta(\sigma - \sigma'). \tag{6.33}$$

Then we have [160]

$$I = \tilde{\kappa} \int d\tau \left(\int d^p \sigma \, \eta_{\mu\nu} \dot{X}^\mu(\tau,\sigma) \dot{X}^\nu(\tau,\sigma) \right)^{1/2}. \qquad (6.34)$$

This is like an action for a point particle in a flat background space,

$$I = m \int d\tau \, (\eta_{\mu\nu} \dot{X}^\mu \dot{X}^\nu)^{1/2}, \qquad (6.35)$$

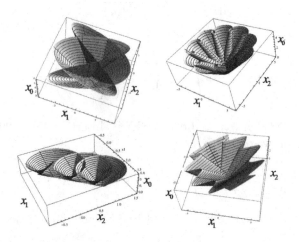

Fig. 6.2   Examples of flat 1-branes for various choices of initial conditions.

but the background space is now infinite dimensional. Variation of (6.34) gives the following equations of motion

$$\frac{d}{d\tau} \left(\frac{\dot{X}^\mu(\tau,\sigma)}{\sqrt{\dot{X}^2}} \right) = 0, \qquad (6.36)$$

where now we have $\dot{X}^2 \equiv \dot{X}^{\nu(\sigma)} \dot{X}_{\nu(\sigma)} = \int d^p \sigma \, \dot{X}^\mu(\sigma) \dot{X}^\nu(\sigma) \eta_{\mu\nu}$. Choosing a gauge in which $\dot{X}^2 = 1$, we obtain the following simple equations of motion:

$$\ddot{X}^\mu(\tau,\sigma) = 0, \qquad (6.37)$$

whose solution is

$$X^\mu(\tau,\sigma) = v^\mu(\sigma)\tau + X_0^\mu(\sigma). \qquad (6.38)$$

This describes a bunch of straight worldlines that altogether form a special kind of brane's worldsheet, namely a worldsheet of a "flat brane".

Equation (6.38) thus describes a continuum limit of a system of non interacting point particles, tracing straight worldlines.

In Fig. 6.2 we give examples of flat 1-branes (i.e., strings) for various solutions of Eq. (6.38), i.e., for various choices of $v^\mu(\sigma)$. We see that flat branes can form involved self-intersecting objects in spacetime. In the last example in Fig. 6.2 the worldsheet does not self intersect, which is a consequence of suitable boundary conditions.

In Fig. 6.3 we illustrate how the situation looks in the case of a metric that differs from (6.33). The chosen metric

$$\rho_{\mu(\sigma)\nu(\sigma')} = \left(1 + \frac{X'^2}{\dot{X}^2}\right)\eta_{\mu\nu}\delta(\sigma - \sigma'), \qquad (6.39)$$

is just that of the usual string, but with different boundary conditions, namely such that self intersections are not excluded.

For comparison, in Fig. 6.4 we show two examples of the usual Nambu-Goto strings for which the metric is (6.39) and the boundary conditions together with the imposed constraints exclude self intersections.

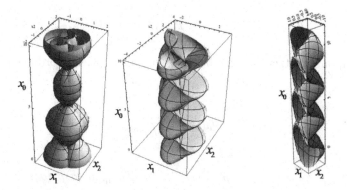

Fig. 6.3   Examples of "curved" 1-branes for various choices of initial conditions. In all cases the brane space metric is $\rho_{\mu(\sigma)\nu(\sigma')} = \left(1 + \frac{X'^2}{\dot{X}^2}\right)\eta_{\mu\nu}\delta(\sigma - \sigma')$.

Quantization of the system described by the action (6.34) can be performed in analogous way as the quantization of the point particle, described by (6.35).

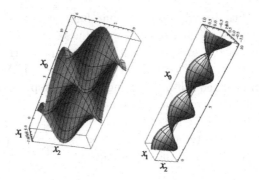

Fig. 6.4 Examples of a special kind of curved 1-branes: the Nambu-Goto strings.

In the case of the point particle (6.35), we have the constraint

$$p^\mu p_\mu - m^2 = 0 , \qquad p_\mu = \frac{m\dot{X}_\mu}{(\dot{X}^2)^{1/2}} , \qquad (6.40)$$

which upon quantization becomes the Klein-Gordon equation,

$$(\hat{p}^\mu \hat{p}_\mu - m^2)\phi(x^\mu) = 0 , \qquad \hat{p}_\mu = -i\partial_\mu \equiv -i\frac{\partial}{\partial x^\mu} . \qquad (6.41)$$

In the case of the brane (6.22) with the metric (6.33) we have the constraint

$$p^{\mu(\sigma)} p_{\mu(\sigma)} - \tilde{\kappa}^2 = 0 , \qquad p_{\mu(\sigma)} = \frac{\tilde{\kappa}\dot{X}_{\mu(\sigma)}}{\sqrt{\dot{X}^2}} , \qquad (6.42)$$

which upon quantization becomes the generalized Klein-Gordon equation,

$$(\hat{p}^{\mu(\sigma)} \hat{p}_{\mu(\sigma)} - \tilde{\kappa}^2)\phi(x^{\nu(\sigma)}) = 0 , \qquad (6.43)$$

$$\hat{p}_{\mu(\sigma)} \equiv -i\partial_{\mu(\sigma)} \equiv -i\frac{\partial}{\partial x^{\mu(\sigma)}} = -i\frac{\delta}{\delta x^\mu(\sigma)} . \qquad (6.44)$$

Here

$$p^{\mu(\sigma)} p_{\mu(\sigma)} = \rho_{\mu(\sigma)\nu(\sigma')}p^{\mu(\sigma)}p^{\nu(\sigma')} = \int \mathrm{d}^p\sigma \, \eta_{\mu\nu}p^\mu(\sigma)p^\nu(\sigma), \qquad (6.45)$$

and

$$\phi(x^{\nu(\sigma)}) \equiv \phi[x^\mu(\sigma)] \qquad (6.46)$$

is a functional of the brane's embedding functions. The point particle equation (6.41) can be derived from the action

$$I[\phi(x^\mu)] = \frac{1}{2}\int \mathrm{d}^4x \, (\partial_\mu\phi\partial^\mu\phi - m^2\phi^2), \qquad (6.47)$$

whereas the corresponding action for the brane equation (6.43) is

$$I[\phi(x^{\mu(\sigma)})] = \frac{1}{2} \int \mathcal{D}x^{\nu(\sigma)} (\partial_{\mu(\sigma)}\phi \, \partial^{\mu(\sigma)}\phi - \tilde{\kappa}^2 \phi^2). \tag{6.48}$$

Explicitly, the equation of motion derived from the latter action is

$$\left(\partial_{\mu(\sigma)}\partial^{\mu(\sigma)} + \tilde{\kappa}^2 \right) \phi = 0. \tag{6.49}$$

In ordinary notation this reads

$$\int d^p\sigma d^p\sigma' \, \eta^{\mu\nu} \delta(\sigma - \sigma') \left(\frac{\delta^2}{\delta x^\mu(\sigma)\delta x^\nu(\sigma')} + \tilde{\kappa}^2 \right) \phi = 0. \tag{6.50}$$

As a classical flat brane is like a bunch of free point particles, so a ("first") quantized brane is like a "bunch", that is, a continuous set of "free", i.e., non interacting quantum fields. Therefore, we can write a solution of Eq. (6.49) as the product [160]

$$\phi(x^{\mu(\sigma)}) = \prod_{\sigma''} \varphi^{(\sigma'')}(x^{\mu(\sigma'')}), \tag{6.51}$$

where for every σ'' we have a field $\varphi^{(\sigma'')}$ which is a function of *four* spacetime coordinates $x^{\mu(\sigma'')}$ that bear a label σ''. This is just like a separation of variables that is commonly used in solutions of partial differential equations. We will now use Eq. (6.18) and introduce the mass

$$m = \kappa\Delta\sigma \tag{6.52}$$

within a region $\Delta\sigma \equiv \Delta^p\sigma \equiv \Delta\sigma^1\Delta\sigma^2...\Delta\sigma^p$. We will also use the following relation between the functional derivative and the partial derivative at a fixed point σ on the brane:

$$\partial_{\mu(\sigma)}\phi \equiv \frac{\delta}{\delta x^\mu(\sigma)}\phi = \lim_{\Delta\sigma \to 0} \frac{1}{\Delta\sigma} \frac{\partial_\mu \varphi^{(\sigma)}}{\partial x^{\mu(\sigma)}} \prod_{\sigma'' \neq \sigma} \varphi^{(\sigma'')}(x^{\mu(\sigma'')}). \tag{6.53}$$

From (6.49), (6.51)–(6.53) we thus obtain

$$\left(\eta^{\mu\nu} \frac{\partial^2}{\partial x^{\mu(\sigma)}\partial x^{\nu(\sigma)}} + m^2 \right) \varphi^{(\sigma)}(x^{\mu(\sigma)}) = 0. \tag{6.54}$$

Because σ is fixed, we can now rename the four spacetime coordinates $x^{\mu(\sigma)}$ into x^μ and write the latter equation simply as

$$\left(\eta^{\mu\nu} \frac{\partial^2}{\partial x^\mu\partial x^\nu} + m^2 \right) \varphi^{(\sigma)}(x^\mu) = 0. \tag{6.55}$$

In our setup a segment of a classical flat brane around σ behaves as a free point particle, and after quantization it satisfies at each σ the Klein-Gordon equation. Because σ is any point on the brane, we have a continuous set of non interacting scalars fields $\varphi^{(\sigma)}$, every one of them satisfying the Klein-Gordon equation (6.55). In other words, we describe the flat brane by means of many particle non interacting field theory. Different segments of the brane behave as distinguishable particles, each being described by a different scalar field.

In the case of a discrete set of non interacting scalar fields $\varphi^r(x)$, the system is described by the action

$$I[\varphi^r(x)] = \frac{1}{2} \int \mathrm{d}^D x \sum_{r=1}^{N} \left(\partial_\mu \varphi^r \partial^\mu \varphi^r - m^2 (\varphi^r)^2 \right). \qquad (6.56)$$

In the continuum limit, the discrete index r becomes the continuous index (σ), and $\varphi^r(x)$ becomes $\varphi^{(\sigma)}(x) \equiv \varphi(\sigma, x)$, or shortly, $\varphi^{(\sigma)} \equiv \varphi(\sigma)$. The action is then

$$I[\varphi(\sigma)] = \frac{1}{2} \int \mathrm{d}^D x \int \mathrm{d}^p \sigma \left(\partial_\mu \varphi(\sigma) \partial^\mu \varphi(\sigma) - m^2 \varphi^2(\sigma) \right). \qquad (6.57)$$

A discrete system based on the action (6.56) can be straightforwardly second quantized, and so can be the continuous system (6.57). In the discrete case, the canonically conjugate variables, the fields $\varphi^r(t, \boldsymbol{x})$ and momenta $\Pi^r(t, \boldsymbol{x})$, become the operators satisfying equal t commutation relations

$$[\varphi^r(t, \boldsymbol{x}), \Pi^s(t, \boldsymbol{x}')] = i \delta^3(\boldsymbol{x} - \boldsymbol{x}') \delta^{rs},$$
$$[\varphi^r(t, \boldsymbol{x}), \varphi^s(t, \boldsymbol{x}')] = 0, \qquad [\Pi^r(t, \boldsymbol{x}), \Pi^s(t, \boldsymbol{x}')] = 0. \qquad (6.58)$$

6.4   An interacting bunch of scalar fields

The action for our system of a continuous set of non interacting scalar fields (6.57) can be written in the form [160]

$$I[\varphi^{(\sigma)}] = \frac{1}{2} \int \mathrm{d}^D x \left(\partial_\mu \varphi^{(\sigma)} \partial^\mu \varphi^{(\sigma')} - m^2 \varphi^{\sigma)} \varphi^{(\sigma')} \right) s_{(\sigma)(\sigma')}, \qquad (6.59)$$

where

$$s_{(\sigma)(\sigma')} = \delta_{(\sigma)(\sigma')} = \delta(\sigma - \sigma'). \qquad (6.60)$$

The latter form of the action suggests its generalization to a continuous set of *interacting* fields. We see that $s_{(\sigma)(\sigma')}$ has the rôle of a metric in the space of the fields $\varphi^{(\sigma)}$. In principle it need not be the simple metric (6.60), but can be a generic metric. In such a way we introduce interactions among the fields, satisfying the action principle (6.59) in which now $s_{(\sigma)(\sigma')}$ is no longer the simple metric (6.60), but a more general metric.

The equation of motion derived from the action (6.59) are

$$\partial_\mu \partial^\mu \varphi_{(\sigma)} + m^2 \varphi_{(\sigma)} = 0, \tag{6.61}$$

where $\varphi_{(\sigma)} = s_{(\sigma)(\sigma')} \varphi^{(\sigma')}$. Assuming that $s_{(\sigma)(\sigma')}$ has the inverse $s^{(\sigma)(\sigma')}$, so that

$$s^{(\sigma)(\sigma'')} s_{(\sigma'')(\sigma')} = \delta^{(\sigma)}{}_{(\sigma')} \equiv \delta(\sigma - \sigma'), \tag{6.62}$$

then we also have $\varphi^{(\sigma)} = s^{(\sigma)(\sigma')} \varphi_{(\sigma')}$. Applying the latter relation on Eq. (6.61), we obtain

$$\partial_\mu \partial^\mu \varphi^{(\sigma)} + m^2 \varphi^{(\sigma)} = 0, \tag{6.63}$$

which is the equation of motion for $\varphi^{(\sigma)}$.

A peculiar property of the system so constructed is that even when the metric is non trivial so that there are interactions among the fields, a general solution of the equation of motion (6.61) has the familiar form

$$\varphi_{(\sigma)}(x) = \int \frac{\mathrm{d}^{\bar{D}} k}{\sqrt{(2\pi)^{\bar{D}} 2\omega_k}} \left(a_{(\sigma)}(k) e^{-ikx} + a^\dagger_{(\sigma)}(k) e^{ikx} \right), \tag{6.64}$$

where $\omega_k = \sqrt{k^2 + m^2}$. The quantities $\varphi_{(\sigma)}(x)$, $a_{(\sigma)}(k)$, and $a^\dagger_{(\sigma)}(k)$ can be raised by means of the inverse metric $s^{(\sigma)(\sigma')}$, so that we obtain

$$\varphi^{(\sigma)}(x) = \int \frac{\mathrm{d}^{\bar{D}} k}{\sqrt{(2\pi)^{\bar{D}} 2\omega_k}} \left(a^{(\sigma)}(k) e^{-ikx} + a^{(\sigma)\dagger}(k) e^{ikx} \right), \tag{6.65}$$

which is a solution of Eq. (6.63).

The canonically conjugated variables $\varphi^{(\sigma)}$ and $\Pi_{(\sigma)} = \partial \mathcal{L}/\partial \dot{\varphi}^{(\sigma)} = \dot{\varphi}_{(\sigma)}$ satisfy

$$[\varphi^{(\sigma)}(x), \Pi_{(\sigma')}(x')] \Big|_{x^0 = x'^0} = \delta^{(\sigma)}{}_{(\sigma')} \delta^{\bar{D}}(x - x'), \tag{6.66}$$

$$[\varphi^{(\sigma)}(x), \varphi^{(\sigma')}(x')] \Big|_{x^0 = x'^0} = 0, \qquad [\Pi_{(\sigma)}(x), \Pi_{(\sigma')}(x')] \Big|_{x^0 = x'^0} = 0. \tag{6.67}$$

From those quantities we construct the Hamiltonian as usual,

$$H = \int \mathrm{d}^{\bar{D}} x \, (\Pi_{(\sigma)} \dot{\varphi}^{(\sigma)} - \mathcal{L}) = \frac{1}{2} \int \mathrm{d}^{\bar{D}} x \, (\Pi_{(\sigma)} \Pi^{(\sigma)} - \partial_i \varphi^{(\sigma)} \partial^i \varphi_{(\sigma)} + m^2 \varphi^{(\sigma)} \varphi_{(\sigma)}), \tag{6.68}$$

and rewrite it in terms the operators $a_{(\sigma)}(\boldsymbol{k})$, and $a^\dagger_{(\sigma)}(\boldsymbol{k})$. From Eqs. (6.64)–(6.67) we find that the latter operators must satisfy

$$[a^{(\sigma)}(\boldsymbol{p}), a^\dagger_{(\sigma')}(\boldsymbol{p}')] = \delta^{(\sigma)}{}_{(\sigma')}\delta^{\bar{D}}(\boldsymbol{p}-\boldsymbol{p}'), \tag{6.69}$$

$$[a^{(\sigma)}(\boldsymbol{p}), a_{(\sigma')}(\boldsymbol{p}')] = 0, \qquad [a^{(\sigma)\dagger}(\boldsymbol{p}), a^\dagger_{(\sigma')}(\boldsymbol{p}')] = 0. \tag{6.70}$$

The relation (6.69) can be written in the following equivalent forms:

$$[a_{(\sigma)}(\boldsymbol{p}), a^\dagger_{(\sigma')}(\boldsymbol{p}')] = s_{(\sigma)(\sigma')}\delta^{\bar{D}}(\boldsymbol{p}-\boldsymbol{p}'), \tag{6.71}$$

$$[a^{(\sigma)}(\boldsymbol{p}), a^{\dagger(\sigma')}(\boldsymbol{p}')] = s^{(\sigma)(\sigma')}\delta^{\bar{D}}(\boldsymbol{p}-\boldsymbol{p}'). \tag{6.72}$$

The Hamilton then becomes

$$H = \frac{1}{2}\int \mathrm{d}^{\bar{D}}\boldsymbol{k}\,\omega_{\boldsymbol{k}}\left(a^\dagger_{(\sigma)}(\boldsymbol{k})a^{(\sigma)}(\boldsymbol{k}) + a^{(\sigma)}(\boldsymbol{k})a^\dagger_{(\sigma)}(\boldsymbol{k})\right)$$

$$= \int \mathrm{d}^{\bar{D}}\boldsymbol{k}\,\omega_{\boldsymbol{k}}\,a^\dagger_{(\sigma)}(\boldsymbol{k})a^{(\sigma)}(\boldsymbol{k}) + H_{\text{z.p.}}, \tag{6.73}$$

where $H_{\text{z.p.}}$ is the "zero point" Hamiltonian, and

$$a^\dagger_{(\sigma)}(\boldsymbol{k})a^{(\sigma)}(\boldsymbol{k}) = a^{(\sigma)\dagger}(\boldsymbol{k})a^{(\sigma')}(\boldsymbol{k})s_{(\sigma)(\sigma')} = a^\dagger_{(\sigma)}(\boldsymbol{k})a_{(\sigma')}(\boldsymbol{k})s^{(\sigma)(\sigma')}. \tag{6.74}$$

More generally, by using the standard field theoretic techniques that involve the Noether theorem, we obtain the stress-energy tensor

$$T^\mu{}_\nu = \frac{\partial\mathcal{L}}{\partial\partial_\mu\varphi^{(\sigma)}}\partial_\nu\varphi^{(\sigma)} - \mathcal{L}\delta^\mu{}_\nu. \tag{6.75}$$

Integrating the latter tensor over a space like hypersurface, we obtain the D-momentum $P_\nu = \int \mathrm{d}\Sigma_\mu T^\mu{}_\nu$. In the reference frame in which the hypersurface has components $\mathrm{d}\Sigma_\mu = (\mathrm{d}\Sigma_0, 0, 0, ..., 0)$ with $\mathrm{d}\Sigma_0 = \mathrm{d}^{\bar{D}}\boldsymbol{x}$, the zero component of the D-momentum is the Hamiltonian (6.68), whilst the spatial components are $P_{\bar{\mu}} = \int \mathrm{d}^{\bar{D}}\boldsymbol{x}\,\dot{\varphi}_{(\sigma)}\partial_{\bar{\mu}}\varphi^{(\sigma)}$, where $\bar{\mu} = 1, 2, ..., \bar{D}$. After using the expansion (6.64), (6.65), we have

$$\hat{\boldsymbol{p}} = \frac{1}{2}\int \mathrm{d}^{\bar{D}}\boldsymbol{k}\,\boldsymbol{k}\left(a^\dagger_{(\sigma)}(\boldsymbol{k})a^{(\sigma)}(\boldsymbol{k}) + a^{(\sigma)}(\boldsymbol{k})a^\dagger_{(\sigma)}(\boldsymbol{k})\right)$$

$$= \int \mathrm{d}^{\bar{D}}\boldsymbol{k}\,\boldsymbol{k}\,a^\dagger_{(\sigma)}(\boldsymbol{k})a^{(\sigma)}(\boldsymbol{k}) + \hat{\boldsymbol{p}}_{\text{z.p.}}, \tag{6.76}$$

where $\hat{\boldsymbol{p}}_{\text{z.p.}}$ is the "zero point" momentum.

By means of the operators $a_{(\sigma)}(\boldsymbol{k})$ and $a^\dagger_{(\sigma)}(\boldsymbol{k})$ we can construct the states of our system. Defining the vacuum state according to

$$a_{(\sigma)}(\boldsymbol{k})|0\rangle = 0, \tag{6.77}$$

the states with definite momenta are created by $a^\dagger_{(\sigma)}(\boldsymbol{k})$,

$$|\boldsymbol{k}_1\rangle = a^\dagger_{(\sigma)}(\boldsymbol{k}_1)|0\rangle \ , \qquad |\boldsymbol{k}_1\boldsymbol{k}_2\rangle = a^\dagger_{(\sigma)}(\boldsymbol{k}_1)a^\dagger_{(\sigma)}(\boldsymbol{k}_2)|0\rangle \ ,.... \qquad (6.78)$$

These are basis states, from which we can form various more general states. For instance, we can form single particle wave packet profile states at every σ, and sum (i.e., integrate) them over σ:

$$|\psi\rangle = \int \mathrm{d}\boldsymbol{p}\, g^{(\sigma)}(\boldsymbol{p})a^\dagger_{(\sigma)}(\boldsymbol{p})|0\rangle, \qquad (6.79)$$

where

$$g^{(\sigma)}(\boldsymbol{p})a_{(\sigma)}(\boldsymbol{p}) = g^{(\sigma)}(\boldsymbol{p})a^{\dagger(\sigma')}(\boldsymbol{p})s_{(\sigma)(\sigma')}. \qquad (6.80)$$

The action of an annihilation operator on such a state gives

$$a^{(\sigma')}(\boldsymbol{p}')|\psi\rangle = g^{(\sigma')}(\boldsymbol{p}')|0\rangle, \qquad (6.81)$$

so that we have

$$\langle\psi|a^{\dagger(\sigma'')}(\boldsymbol{p}'')a^{(\sigma')}(\boldsymbol{p}')|\psi\rangle = g^{*(\sigma'')}(\boldsymbol{p}'')g^{(\sigma')}(\boldsymbol{p}')\langle 0|0\rangle. \qquad (6.82)$$

We normalize the vacuum according to $\langle 0|0\rangle = 1$.

Let us now consider the state which is the product of "single particle" wave packet profiles [160]:

$$|\psi\rangle = \prod_\sigma \int \mathrm{d}^{\bar{D}}\boldsymbol{p}_{(\sigma)}\, g^{(\sigma)}(\boldsymbol{p}_{(\sigma)})a^\dagger_{(\sigma)}(\boldsymbol{p}_{(\sigma)})|0\rangle \qquad \text{no integration over } (\sigma).$$
$$(6.83)$$

The action of an annihilation operator to the latter state gives

$$a^{(\sigma')}(\boldsymbol{p}'_{(\sigma')})|\psi\rangle = \int \mathrm{d}\boldsymbol{p}_{(\sigma)}\mathrm{d}\sigma'\delta(\sigma' - \sigma)\delta(\boldsymbol{p}'_{(\sigma')} - \boldsymbol{p}_{(\sigma)})g^{(\sigma)}(\boldsymbol{p}_{(\sigma')})|\bar\psi\rangle$$

$$= g^{(\sigma')}(\boldsymbol{p}'_{(\sigma')})|\bar\psi\rangle, \qquad (6.84)$$

where $|\bar\psi\rangle$ is the product of all the single "particle" states. except the one picked up by $a^{(\sigma')}(\boldsymbol{p}'_{(\sigma')})$:

$$|\bar\psi\rangle = \left(\prod_{\sigma\neq\sigma'} \int \mathrm{d}\boldsymbol{p}_{(\sigma)}g^{(\sigma)}(\boldsymbol{p}_{(\sigma)})a^\dagger_{(\sigma)}(\boldsymbol{p}_{(\sigma)})\right)|0\rangle. \qquad (6.85)$$

We thus have

$$\langle\psi|a^{\dagger(\sigma'')}(\boldsymbol{p}''_{(\sigma'')})a^{(\sigma')}(\boldsymbol{p}'_{(\sigma')})|\psi\rangle = g^{*(\sigma'')}(\boldsymbol{p}''_{(\sigma'')})g^{(\sigma')}(\boldsymbol{p}'_{(\sigma')})\langle\bar\psi|\bar\psi\rangle, \qquad (6.86)$$

where normalization can be such that $\langle\bar\psi|\bar\psi\rangle = 1$.

We are now going to calculate how the expectation value of the momentum operator changes with time. Using the Schrödinger equation we obtain [160]

$$\frac{d}{dt}\langle\psi|\hat{p}|\psi\rangle = \left(\frac{d}{dt}\langle\psi|\right)\hat{p}|\psi\rangle + \langle\psi|\hat{p}\frac{d}{dt}|\psi\rangle = (-i)\langle\psi|\hat{p}H - H^\dagger\hat{p}|\psi\rangle. \quad (6.87)$$

In the above derivation we assumed that the Hamilton operator is not Hermitian. This is the case, if the mass $m = \kappa\Delta\sigma$ depends on position σ on the brane[2], so that also $\sqrt{m^2(\sigma) + k^2} = \omega_k(\sigma)$ is a function of σ. From the expression (6.73) for the Hamiltonian in which instead of a σ independent ω_k stays $\omega_k(\sigma)$, we then find $H^\dagger \neq H$.

If we insert into Eq. (6.87) either a state (6.73) or (6.83) we obtain [160]

$$\frac{d}{dt}\langle\psi|\hat{p}|\psi\rangle = (-i)\int d^{\bar{D}}p\, p\, g^*(\sigma, p)g(\sigma', p)s(\sigma, \sigma')(\omega_p(\sigma) - \omega_p(\sigma'))d\sigma d\sigma', \quad (6.88)$$

where we now write $g^{(\sigma)}(p) \equiv g(\sigma, p)$, and $s_{(\sigma)(\sigma')} \equiv s(\sigma, \sigma')$ and explicitly denote the integration over σ and σ' In the case of a σ independent ω_p the above expression vanishes, which means that the expectation values of the system's total momentum is conserved in time. This is indeed the case for an isolated system, whose tension κ, and thus ω_k cannot change with σ. If the system is in interaction with another system, then in principle tension can depend on σ.

Let us now assume that there is the following local interaction between nearby brane segments [160]:

$$s_{(\sigma)(\sigma')} \equiv s(\sigma, \sigma') = (1 + \lambda\partial_{\bar{a}}\partial^{\bar{a}})\delta^p(\sigma - \sigma'). \quad (6.89)$$

Using the latter expression in Eq. (6.88), we obtain

$$\frac{d}{dt}\langle p\rangle \equiv \frac{d}{dt}\langle\psi|\hat{p}|\psi\rangle = (-i)\lambda\int d^{\bar{D}}p\, d^p\sigma\, p\,\omega_p(\sigma)(g^*\partial_{\bar{a}}\partial^{\bar{a}}g - \partial_{\bar{a}}\partial^{\bar{a}}g^*g). \quad (6.90)$$

In the expression for the total brane's momentum,

$$\langle\psi|\hat{p}|\psi\rangle = \int dp\, d\sigma\, d\sigma'\, p\, g^*(\sigma, p)g(\sigma', p)s(\sigma, \sigma')$$

$$= \int dp\, d\sigma\, p(g^*g + g^*\partial_{\bar{a}}\partial^{\bar{a}}g) = \langle\hat{p}\rangle = \int d\sigma\langle\hat{p}\rangle_\sigma \quad (6.91)$$

there is the integrations over $d\sigma \equiv d^p\sigma$. If we omit this integration, then we have the expected momentum density of a brane's segment:

$$\langle\hat{p}\rangle_\sigma = \int dp\, p(g^*g + g^*\partial_{\bar{a}}\partial^{\bar{a}}g) = \langle\psi|_\sigma\hat{p}|\psi\rangle_\sigma, \quad (6.92)$$

---

[2]In the discrete case this is equivalent to every particle (field) having a different mass m_r. In the continuous case this means that the brane's tension κ is σ dependent.

where

$$|\psi\rangle_\sigma = \int \mathrm{d}\boldsymbol{p}\, g(\sigma, \boldsymbol{p}) a^\dagger(\sigma, \boldsymbol{p})|0\rangle, \tag{6.93}$$

is the state of the brane's element at $\sigma' \equiv \sigma'^{\bar{a}}$, i.e., the state (6.83) with the product over σ being omitted.

The time derivative of such an expected momentum density is obtained from Eq. (6.90), if we omit the integrations over $\mathrm{d}^P\sigma$:

$$\frac{\mathrm{d}}{\mathrm{d}t}\langle\hat{\boldsymbol{p}}\rangle_\sigma = (-i)\lambda \int \mathrm{d}^{\bar{D}}\boldsymbol{p}\, \boldsymbol{p}\, \omega_{\boldsymbol{p}}(g^*\partial_{\bar{a}}\partial^{\bar{a}}g - \partial_{\bar{a}}\partial^{\bar{a}}g^*g). \tag{6.94}$$

The latter expression can be different from zero even if $\omega_{\boldsymbol{p}}$ does not change with σ. In fact this is the continuity equation for the current density on the brane, isolated from its environment.

If, instead a wave packet profile in momentum space, we take a wave packet in coordinate space, the Fourier transformation being

$$g(\sigma, \boldsymbol{p}) = \frac{1}{(2\pi)^{\bar{D}/2}} \int \mathrm{e}^{-i\boldsymbol{p}\boldsymbol{x}} f(\sigma, \boldsymbol{x}) \mathrm{d}\boldsymbol{x}, \tag{6.95}$$

then Eq. (6.94) becomes

$$\frac{\mathrm{d}}{\mathrm{d}t}\langle\hat{\boldsymbol{p}}\rangle_\sigma = -\lambda\partial_{\bar{a}} \int \mathrm{d}^{\bar{D}}\boldsymbol{x} \left[f^*(\sigma, \boldsymbol{x}) \left(\nabla\sqrt{m^2 + (-i\nabla)^2}\, \partial^{\bar{a}} f(\sigma, \boldsymbol{x}) \right) \right.$$
$$\left. - \left(\nabla\sqrt{m^2 + (-i\nabla)^2}\, \partial^{\bar{a}} f^*(\sigma, \boldsymbol{x}) \right) f(\sigma, \boldsymbol{x}) \right]. \tag{6.96}$$

Though we have not explicitly denoted so, wave packet profiles $g(\sigma, \boldsymbol{p})$ and $f(\sigma, \boldsymbol{x})$ depend on time as well. Therefore a state such as (6.93) is time dependent and satisfies the time dependent Schrödinger equation with the Hamilton operator (6.73). The wave packet profile then satisfies [162–164], [165]

$$\sqrt{m^2 + (-i\nabla^2)}\, f = -i\frac{\partial}{\partial t} f. \tag{6.97}$$

Using the latter equation, we can express (6.96) in terms of the time derivative:

$$\frac{\mathrm{d}}{\mathrm{d}t}\langle\hat{\boldsymbol{p}}\rangle_\sigma = -\lambda\partial_{\bar{a}} \int \mathrm{d}^{\bar{D}}\boldsymbol{x} \left[f^*(\sigma, \boldsymbol{x}) \left(\nabla(-i)\frac{\partial}{\partial t}\partial^{\bar{a}} f(\sigma, \boldsymbol{x}) \right) \right.$$
$$\left. - \left(\nabla(-i)\frac{\partial}{\partial t}\partial^{\bar{a}} f^*(\sigma, \boldsymbol{x}) \right) f(\sigma, \boldsymbol{x}) \right]. \tag{6.98}$$

If we rewrite Eq. (6.98) in components,

$$\frac{\mathrm{d}}{\mathrm{d}t}\langle\hat{p}_{\bar{\mu}}\rangle_\sigma = -\lambda\partial_{\bar{a}} \int \mathrm{d}^{\bar{D}}\boldsymbol{x} \left[f^*(\sigma, \boldsymbol{x}) \left(-i\frac{\partial}{\partial t}\partial^{\bar{a}}\partial_{\bar{\mu}} f \right) - \left(-i\frac{\partial}{\partial t}\partial^{\bar{a}}\partial_{\bar{\mu}} f^* \right) f \right], \tag{6.99}$$

where $\nabla \equiv \partial_{\bar{\mu}}$, $\bar{\mu} = 1, 2, ...\bar{D}$, then we immediately recognize that the right-hand side of Eq. (6.99) is the divergence of the expectation value of the operator

$$\hat{\pi}^{\bar{a}}{}_{\bar{\mu}} = -i\lambda \frac{\partial}{\partial t} \overset{\leftrightarrow}{\partial}{}^{\bar{a}} \partial_{\bar{\mu}} \equiv -i\lambda \frac{\partial}{\partial t} \left(\overset{\leftarrow}{\partial}{}^{\bar{a}} - \overset{\rightarrow}{\partial}{}^{\bar{a}} \right) \partial_{\bar{\mu}}. \qquad (6.100)$$

Close to the initial time $t = 0$ the solution of Eq. (6.97) for a minimal wave packet can be approximated with a Gaussian wave packet if its width is greater than the Compton wavelength:

$$f \approx A e^{-\frac{(\boldsymbol{x} - \bar{\boldsymbol{X}}(\sigma))^2}{2\bar{\sigma}_0}} e^{i\bar{\boldsymbol{p}}\boldsymbol{x}} e^{i\bar{p}_0 t}, \qquad (6.101)$$

where $\bar{\boldsymbol{X}}(\sigma)$, $\bar{\boldsymbol{p}}$ and $\bar{p}_0$ are the coordinates, momentum and energy of the wave packet center, respectively, whilst A is the normalization constant.

Inserting the wave packet (6.101) into (6.99), we obtain [160]

$$\frac{\langle \hat{p}_{\bar{\mu}} \rangle_\sigma}{dt} = -\lambda \partial_{\bar{a}} \left(\frac{\bar{p}_0}{\Delta S} \frac{\partial^{\bar{a}} \bar{X}_{\bar{\mu}}}{\sigma_o} \right), \qquad (6.102)$$

where $\Delta S = \int d^p \sigma$. This is reminiscent of the brane equation of motion (6.32).

Let us now consider the following metric in the field space, covariant under reparametrizations of the brane parameters $\sigma^{\bar{a}}$:

$$s(\sigma, \sigma') = \sqrt{-\bar{\gamma}(\sigma)} \, \delta^p(\sigma - \sigma') + \partial_{\bar{a}} \left(\sqrt{-\bar{\gamma}(\sigma)} \gamma^{\bar{a}\bar{b}} \partial_{\bar{b}} \right) \delta^p(\sigma - \sigma') \quad (6.103)$$

where $\gamma \equiv \det\gamma_{\bar{a}\bar{b}}$. With such a metric, instead of (6.102) we obtain [160]

$$\frac{\langle \hat{p}_{\bar{\mu}} \rangle_\sigma}{dt} = -\lambda \partial_{\bar{a}} \left(\frac{\bar{p}_0}{\Delta S} \frac{\sqrt{-\bar{\gamma}} \gamma^{\bar{a}\bar{b}} \partial_{\bar{b}} \bar{X}_{\bar{\mu}}}{\sigma_o} \right), \qquad (6.104)$$

where now we have $\Delta S = \int \sqrt{-\bar{\gamma}(\sigma)}$. The latter equation is in fact the equation of motion (6.32) of a classical Dirac-Nambu-Goto brane if we make the following correspondence:

$$\langle \hat{p}_{\bar{\mu}} \rangle_\sigma \equiv \langle \hat{\boldsymbol{p}} \rangle_\sigma \longrightarrow p_{\bar{\mu}}(\sigma) = \frac{\kappa \sqrt{-\bar{\gamma}} \dot{X}_{\bar{\mu}}}{\sqrt{\dot{X}^2}} \qquad (6.105)$$

$$\frac{\bar{p}_0}{\Delta S} \gamma^{\bar{a}\bar{b}} \partial_{\bar{b}} \bar{X}_{\bar{\mu}} \longrightarrow \kappa \sqrt{-\bar{\gamma}} \sqrt{\dot{X}^2} \partial^{\bar{a}} X_{\bar{\mu}} = p_0(\sigma) \dot{X}^2 \partial^{\bar{a}} X_{\bar{\mu}}, \qquad (6.106)$$

and take $\dot{X}^2 = 1 - v^2 = 1$. The latter equality holds in a gauge in which $\tau = x^0$, if $v^2 = 0$. Recall that Eq. (6.104) has been calculated for the wave packet at $t \approx 0$, therefore $v^2 = 0$ is consistent with vanishing $\langle \hat{p}_{\bar{\mu}} \rangle \propto \dot{\bar{X}}_{\bar{\mu}}$, $\bar{\mu} = 1, 2, ..., \bar{D}$ at $t \approx 0$.

6.5   Generalization to arbitrary configurations

The exercises with the brane space were just a tip of an iceberg. Instead of one brane, a configuration can consist of many branes, or point particles, or both, as illustrated in Fig. 6.5. The action for such a system is a straightforward generalization of the brane action (6.22) to such an extended configurations space C:

$$I = \tilde{\kappa} \int d\tau (\rho_{MN} \dot{X}^M \dot{X}^N)^{1/2}. \tag{6.107}$$

Here we use the same compact indices M, N for coordinates in C in various cases:

$$
\begin{aligned}
M &= \mu i && \text{many point particles} \\
M &= \mu(\sigma) && \text{a single brane} \\
M &= k\mu(\sigma) && \text{many branes} \\
M &= \mu_1 \mu_2 ... \mu_r && \text{oriented } r\text{-volume associated with a brane,}
\end{aligned}
$$

where $\mu = 0, 1, 2, ..., \bar{D}$ denotes coordinates of D-dimensional spacetime, $i = 1, 2, 3, ...$ counts different particles, and $k = 1, 2, 3, ...$ different branes. The meaning of the last line will be explained shortly below.

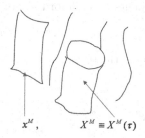

$$x^M, \qquad X^M \equiv X^M(\tau)$$

Fig. 6.5   A configuration can consist of many branes, or point particles, or both.

We thus adopt a generic notation so that x^M and $X^M \equiv X^M(\tau)$ denotes, respectively, coordinates and τ-dependent functions in whatever configuration space, either a system of many particles, a single brane, or a system of many branes, or a Clifford space associated with a brane. Thus, depending on the considered physical system, $M = \mu i$, $M = \mu(\sigma)$, $M = k\mu(\sigma)$, or $M = \mu_1 \mu_2 ... \mu_r$. Then Eq. (6.107) and derived equations are valid for all those cases of configuration spaces.

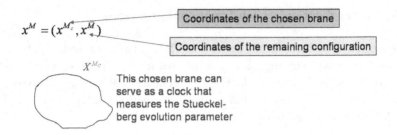

$$x^M = (x^{M_c}, x_\star^M)$$

Coordinates of the chosen brane

Coordinates of the remaining configuration

X^{M_c}

This chosen brane can
serve as a clock that
measures the Stueckel-
berg evolution parameter

Fig. 6.6   One of the branes within a configuration can be chosen to serve as a clock.

As a consequence of the invariance of the action (6.107) under reparametrizations of τ, the momenta $p^M = \dfrac{\tilde{\kappa}\dot{X}^M}{\sqrt{\dot{X}^N \dot{X}_N}}$ satisfy the constraint

$$p^M p_M - \tilde{\kappa}^2 = 0. \tag{6.108}$$

Let us consider a configuration which consists of many particles and/or branes. Let us choose one brane and denote its coordinates as X^{M_c} (Fig. 6.6).

A way to sample a brane is to describe it as a set of 16 oriented r-areas (or r-volumes) of all popssible dimensionalities, $r = 0, 1, 2, ..., D$. We shall take $D = 4$. In Refs. [104] it has been shown how a brane, described by an infinite dimensional vector $x^{\mu(\sigma)}$ can be mapped into a vector of the space spanned by the basis elements of a Clifford algebra $Cl(1,3)$:

$$x^{\mu(\sigma)} \to x^{\mu_1 \mu_2 \cdots \mu_r} \gamma_{\mu_a} \wedge \gamma_{\mu_2} \wedge ... \wedge \gamma_{\mu_r} \equiv x^{M_c} \gamma_{M_c}. \tag{6.109}$$

To avoid multiple counting of the terms, it is convenient to order the indices according to $\mu_1 < \mu_2 < ... < \mu_r$, $r = 0, 1, 2, 3, 4$.

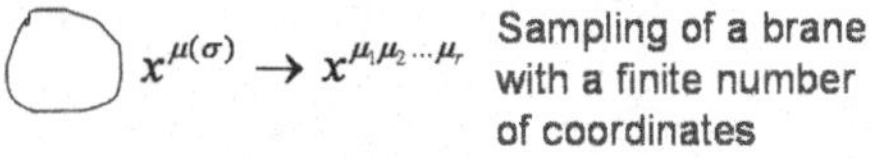

$x^{\mu(\sigma)} \to x^{\mu_1 \mu_2 \cdots \mu_r}$ Sampling of a brane
with a finite number
of coordinates

Fig. 6.7   A brane can be sampled by coordinates of Clifford space.

If instead of one brane we consider two, three or more branes, such a system can also be described by 16 coordinates of the Clifford space [104] (see Fig. 6.8).

Clifford algebra can be considered [49, 99, 100, 104–106] as a tangent space to a manifold, called *Clifford space*, C. We will consider flat Clifford space, which is isomorphic to $Cl(1,3)$. Therefore, the points of C can be described by $x^{\mu_1 \cdots \mu_r}$. In Eq. (6.109) we have thus a mapping from the infinite dimensional brane space to the 16-dimensional Clifford space.

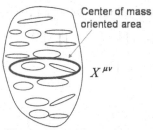

Center of mass
oriented area

$X^{\mu\nu}$

The constituent branes can
have a preferred orientation

Fig. 6.8   An effective ("center of mass") brane associated with a system of many branes.

A brane can be sampled by 16 coordinates $x^M \equiv x^{\mu_1\cdots\mu_r}$ of the Clifford space.

The metric of Clifford space is given by the scalar product of two basis elements:

$$\eta_{M_c N_c} = \gamma^{\ddagger}_{M_c} * \gamma_{N_c} = \langle \gamma^{\ddagger}_{M_c} \gamma_{N_c} \rangle_0, \qquad (6.110)$$

where "$\ddagger$" denotes reversion of the order of vectors in the product $\gamma_{M_c} = \gamma_{\mu_1}\gamma_{\mu_2}\cdots\gamma_{\mu_r}$. The subscript "0" denotes the scalar part of the expression. Explicitly the metric (6.41) is [47]

$$\eta_{M_c N_c} = \mathrm{diag}(1,1,1,1,1,1,1,1,-1,-1,-1,-1,-1,-1,-1,-1). \quad (6.111)$$

Clifford space is thus an ultrahyperbolic space.

The scalar product of $X^{\ddagger} = (x^{M_c}\gamma_{M_c})^{\ddagger}$ and $X = x^{M_c}\gamma_{M_c})$ gives

$$
\begin{aligned}
X^{\ddagger} * X &= \eta_{M_c N_c} x^{M_c} x^{N_c} \\
&= s^2 + \eta_{\mu\nu}x^{\mu}x^{\nu} + \tfrac{1}{4}(\eta_{\mu\beta}\eta_{\nu\alpha} - \eta_{\mu\alpha}\eta_{\nu\beta})x^{\mu\alpha}x^{\nu\beta} + \eta_{\mu\nu}\tilde{x}^{\mu}\tilde{x}^{\nu} - \tilde{s}^2 \\
&= \eta_{\hat{\mu}\hat{\nu}}x^{\hat{\mu}}x^{\hat{\nu}} + s^2 - \tilde{s}^2, \qquad (6.112)
\end{aligned}
$$

where s, $\tilde{s} = \tfrac{1}{4!}\epsilon_{\mu\nu\rho\sigma}x^{\mu\nu\rho\sigma}$ and $\tilde{x}^{\mu} = \tfrac{1}{3!}\epsilon^{\mu}{}_{\nu\rho\sigma}x^{\nu\rho\sigma}$ are the scalar, pseudoscalar and pseudovector coordinates, respectively. In the last expression we introduced $x^{\hat{\mu}} = (x^{\mu}, x^{\mu\nu}, \tilde{x}^{\mu})$. We thus have $x^{M_c} = (s, \tilde{s}, x^{\hat{\mu}})$.

Upon ("first") quantization the constraint (6.108), associated with the action (6.107), becomes the Klein-Gordon equation in the configuration space:

$$\left(\rho^{MN}\frac{\partial^2}{\partial_M \partial_N} + \tilde{\kappa}^2\right)\phi. \qquad (6.113)$$

The corresponding action for the scalar field $\phi(x^M)$ is

$$I = \int \mathcal{D}x^M \left(\rho^{MN}\frac{\partial\phi^*}{\partial x^M}\frac{\partial\phi}{\partial x^N} - \tilde{\kappa}^2\phi^*\phi\right), \qquad (6.114)$$

where $\mathcal{D}x^M \equiv \prod_M dx^M$ is a volume element in the configuration space. In the case of many branes, $\prod_M dx^M = \prod_{k\mu(\sigma)} dx^{k\mu(\sigma)}$, whereas in the case of many particles it is $\prod_M dx^M = \prod_{\mu i} dx^{\mu i}$.

6.5.1 *Non interacting case*

If the metric ρ^{MN} is a generalization of the Minkowski metric to the configuration space, then we have the Klein-Gordon equation in flat configuration space. We will now consider such a non interacting case.

By splitting the index M according to $M = (M_c, \bar{M})$, where M_c refers to one chosen brane, described in terms of the coordinates $x^{M_c} = (s, \tilde{s}, x^{\hat{\mu}})$ of the Clifford space, whereas $\bar{M}$ refers to the remaining particle and/or branes, and then renaming $\bar{M}$ back into M, the field action (6.114) becomes

$$I = \frac{1}{2} \int \mathcal{D}x^{M_c}\, \mathcal{D}x^M \left(\frac{\partial \phi^*}{\partial x^{M_c}} \frac{\partial \phi}{\partial x_{M_c}} + \frac{\partial \phi^*}{\partial x^M} \frac{\partial \phi}{\partial x_M} - \kappa^2 \phi^* \phi \right). \qquad (6.115)$$

Let us introduce the light-cone coordinates

$$\tau = \frac{1}{\sqrt{2}}(s + \tilde{s}), \qquad \lambda = \frac{1}{\sqrt{2}}(s - \tilde{s}), \qquad (6.116)$$

so that instead of the coordinates $x^{M_c} = (s, \tilde{s}, x^{\hat{\mu}})$, we have now the coordinates $x'^{M_c} = (\tau, \lambda, x^{\hat{\mu}})$. The field ϕ then depends on the light-cone coordinates τ, λ, the remaining 12 coordinates $x^{\hat{\mu}}$ of the Clifford space associated with the chosen brane, and on the coordinates x^M of the remaining objects (branes or particles) within the configurations.

Taking the ansatz

$$\phi(\tau, \lambda, x^{\hat{\mu}}, x^M) = e^{i\Lambda\lambda} e^{ip_{\hat{\mu}} x^{\hat{\mu}}} \psi(\tau, x^M), \qquad (6.117)$$

the action (6.115) becomes

$$I_0 = \frac{1}{2} \int d\tau\, \mathcal{D}x^M \left[i\lambda \left(\frac{\partial \psi^*}{\partial \tau} \psi - \psi^* \frac{\partial \psi}{\partial \tau} \right) + \partial_M \psi^* \partial^M \psi - (\kappa^2 - p_{\hat{\mu}} p^{\hat{\mu}}) \psi^* \psi \right].$$
$$(6.118)$$

We have omitted the integration over λ and $x^{\hat{\mu}}$, because it gives a constant factor which can be absorbed into the redefinition of the action I_0. The equation of motion is the Stueckelberg equation in the configuration space:

$$i\frac{\partial \psi}{\partial \tau} = -\frac{1}{2\Lambda}(\partial_M \partial^M + M'^2)\psi, \qquad (6.119)$$

where $M'^2 = \kappa^2 - p_{\hat{\mu}}p^{\hat{\mu}}$. The general solution is

$$\psi(\tau, x^M) = \int Dp^M \, c(p^M) \exp\left[ip_M x^M - \frac{i}{2\Lambda}(p_M p^M - M'^2)\tau\right], \quad (6.120)$$

in which there is no restriction on momenta p_M, therefore initial data at $\tau = 0$ can be freely specified.

In particular it can be $c(p^M) = c(p^{1\mu(\sigma)})c(p^{2\mu(\sigma)})...c(p^{N\mu(\sigma)})$, for a multi brane configuration, or $c(p^M) = c(p^{1\mu})c(p^{2\mu})...c(p^{N\mu})$ for a multi particle configuration. Then the field $\phi(x^M)$ can be written as the product of snigle brane or single particle states. In the case of particles we have:

$$\psi(\tau, x^M) = \varphi(\tau, x^{1\mu})\varphi(\tau, x^{2\mu})...\varphi(\tau, x^{N\mu}) \, e^{\frac{i}{2\Lambda}M'^2\tau}, \quad (6.121)$$

where

$$\varphi(\tau, x^{1\mu}) = \int d^4p_1 \, c(p^{1\mu}) \exp\left[ip_{1\mu}x^{1\mu} - \frac{i}{2\Lambda}p_{1\mu}p^{1\mu}\tau\right], \quad (6.122)$$

and similarly for other particles labeled by $2, 3, ..., N$.

Writing now

$$\psi(\tau, x^M) = \varphi(\tau, x^\mu)\chi(x^{\bar{M}}), \quad (6.123)$$

where $\varphi(\tau, x^\mu) \equiv \varphi(\tau, x^{1\mu})$ is the field associated with a chosen particle (labeled by '1'), and

$$\chi(\tau, x^{\bar{M}}) = \int d\tau \, d^{D-4}\bar{p} \, c(\bar{p}) \, e^{ip_{\bar{M}}x^{\bar{M}}} e^{-\frac{i}{2\Lambda}(p_{\bar{M}}p^{\bar{M}} - M'^2)\tau} \quad (6.124)$$

is the field over the configuration of the remaining particles with coordinates $x^{\bar{M}} \equiv x^{\bar{i}\mu}$, $\bar{i} = 2, 3, ..., N$, the action (6.118) becomes

$$I = Q \int d\tau \, d^4x \left[-i2\Lambda\varphi^* \frac{\partial \varphi}{\partial \tau} + \partial_\mu \varphi^* \partial^\mu \varphi - m^2 \varphi^* \varphi\right], \quad (6.125)$$

where

$$m^2 = \frac{(1-Q)}{Q}M'^2, \quad (6.126)$$

and

$$Q = \int d^{D-4}\bar{x} \, \chi^* \chi = \int d^{D-4}\bar{p} \, c^*(\bar{p})c(\bar{p}). \quad (6.127)$$

We can normalize χ so that $Q = 1$. Then

$$I = \int d^4x \left[-i2\Lambda\varphi^* \frac{\partial \varphi}{\partial \tau} + \partial_\mu \varphi^* \partial^\mu \varphi\right], \quad (6.128)$$

which is the Stueckelberg action [49, 112–122] for a single particle field $\varphi(\tau, x^\mu)$. From (6.128) we obtain the Stueckelberg field equation

$$i\frac{\partial\varphi}{\partial\tau} = -\frac{1}{2\Lambda}\partial_\mu\partial^\mu\varphi. \tag{6.129}$$

The non interacting many particle Stueckelberg equation (6.119) thus contains the single particle Stueckelberg equation (6.129).

Upon quantization, $\varphi(\tau, x^\mu)$ becomes the operator that annihilates, and $\varphi^*(\tau, x^\mu)$ the operator $\varphi^\dagger(\tau, x^\mu)$ that creates a particle (more precisely, an 'instantonic' particle or and 'event') at x^μ. The evolution of the system is given in terms of the Stueckelberg evolution parameter τ, which in our setup is associated with the brane sampled by the coordinates x^{M_c} of the Clifford space. The latter brane[3] is a part of the overall considered configuration, and is given the role of a clock, which can be a "Stueckelberg clock". The Stueckelberg evolution parameter τ is thus embedded in the configuration.

In the Stueckelberg quantum field theory, the position operator is not considered as problematic[4]. It creates an *event* in spacetime.

6.5.2 *Bunch of Stueckelberg fields interacting in a particular way*

The procedure with branes and interacting quantized fields that we have performed in Sec. 6.4 can be done à la Stueckelberg as well. The Stueckelberg field action (6.128) or its more general form (6.125) refers to a single quantum field. Instead of one such a field we can have many fields, and even a continuous set of such fields, as in Sec. 6.4. But instead of the field action (6.59) we now have (upon quantization) the following action

$$I[\varphi^{(\xi)}] = \frac{1}{2}\int d\tau d^D x \left(-i2\Lambda\varphi^{\dagger(\xi)}\frac{\partial\varphi^{(\xi')}}{\partial\tau} + \partial_\mu\varphi^{\dagger(\xi)}\partial^\mu\varphi^{(\xi')}\right.$$

$$\left. -m^2\varphi^{\dagger(\xi)}\varphi^{(\xi')}\right)s_{(\xi)(\xi')}, \tag{6.130}$$

where $\xi \equiv \xi^a$, $a = 1, 2, ..., d$, are d parameters. If the metric is $s_{(\xi)(\xi')} = \delta(\xi, \xi')$, then this is the action for a continuous set of non interacting Stueckelberg fields, otherwise it is an action for interacting Stueckelberg fields. The momentum, canonically conjugate to the field $\varphi^{(\xi)}(x)$ is

---

[3]It need not be only one brane, there can be many branes, altogether sampled by x^{M_c} (see Ref. [104]).

[4]A careful analysis reveals that position operator is not problematic [160] even in the usual quantum field theory.

$\Pi^{(\xi)} = -i\Lambda\varphi^{\dagger(\xi)}(x)$. We have the following commutation relations

$$[\varphi^{(\xi)}(\tau, x), \Pi_{(\xi')}(\tau, x')] = i\delta^{(\xi)}{}_{(\xi')}\delta^D(x - x'), \tag{6.131}$$

$$[\varphi^{(\xi)}(\tau, x), \varphi^{(\xi')}(\tau, x')] = 0, \qquad [\Pi^{(\xi)}(\tau, x), \Pi^{(\xi')}(\tau, x')] = 0. \tag{6.132}$$

The equation of motion derived from (6.130) is

$$i\frac{\partial\varphi_{(\xi')}}{\partial\tau} = -\frac{1}{2\Lambda}(\partial_\mu\partial^\mu + m^2)\varphi_{(\xi')}. \tag{6.133}$$

Its solution can be expanded according to

$$\varphi_{(\xi)}(\tau, x) = \frac{1}{\sqrt{(2\pi)^D\Lambda}}\int \mathrm{d}^D p\, a_{(\xi)}(p)\exp\left[ip_\mu x^\mu + \frac{i}{\Lambda}(p^\mu p_\mu - m^2)\tau\right], \tag{6.134}$$

where the commutation relations (6.131) are satisfied provided that

$$[a^{(\xi)}(p), a^{\dagger}_{(\xi')}(p')] = \delta^{(\xi)}{}_{\xi'}\delta^D(p - p'), \tag{6.135}$$

while, as usually, the commutators of equal type operators, vanish.

An operator $a^{\dagger}_{(\xi)}(p)$ creates and $a^{(\xi)}(p)$ annihilates a (ξ)-type particle with momentum $p \equiv p^\mu$, $\mu = 0, 1, 2, ..., D$. Vacuum state is defined according to $a^{(\xi)}(p)|0\rangle = 0$. The Fourier transformed operators

$$a^{\dagger}_{(\xi)}(x) = \frac{1}{\sqrt{(2\pi)^D}}\int \mathrm{d}^D p\, a^{\dagger}_{(\xi)}(p)e^{-ip_\mu x^\mu}, \tag{6.136}$$

$$a_{(\xi)}(x) = \frac{1}{\sqrt{(2\pi)^D}}\int \mathrm{d}^D p\, a_{(\xi)}(p)e^{ip_\mu x^\mu}, \tag{6.137}$$

are creation and annihilation operators for a particle event at a spacetime point x^μ. Up to a factor $\sqrt{\Lambda}$ they coincide with the field operators $\varphi_{(\xi)}(\tau, x)$ and $\varphi^{\dagger}_{(\xi)}(\tau, x)$ at a fixed value of τ (say $\tau = 0$).

A many particle event state is obtained by successive action of creation operators on the vacuum. In the limit of infinitely many densely packed events such a configuration can be a brane (an extended event) in spacetime:

$$\prod_\xi a^{\dagger}_{(\xi)}(x_\xi)|0\rangle \equiv A^{\dagger}[X^\mu(\xi)]|0\rangle = |X^\mu(\xi)\rangle, \tag{6.138}$$

where $X^\mu(\xi)$ are a brane's embedding functions of d parameters $\xi \equiv \xi^a$, which now need not be all space like; one of them can be time like [49]. In such a case $X^\mu(\xi)$ describes a brane that extends into $d - 1$ spacelike directions and into one time like direction of the embedding space. General states are superposition of the states (6.138) or their momentum space counterparts.

The Hamilton operator is

$$H = \int d^D x (\Pi_{(\xi)} \partial_\tau \varphi^{(\xi)} - \mathcal{L})$$

$$= \frac{1}{2\Lambda} \int d^D x \, (\partial_\mu \varphi^{\dagger(\xi)} \partial^\mu \varphi^{(\xi')} - m^2 \varphi^{\dagger(\xi)} \partial^\mu \varphi^{(\xi')}) s_{(\xi)(\xi')}$$

$$= \frac{1}{2\Lambda} \int d^D p \, (p^2 - m^2) \, a^{\dagger(\xi)}(p) a^{(\xi')}(p) s_{(\xi)(\xi')}. \tag{6.139}$$

Similarly, we obtain the momentum operator:

$$\hat{p}_\mu = \int d^D p \, p_\mu \, a^{\dagger(\xi)}(p) a^{(\xi')}(p) s_{(\xi)(\xi')}. \tag{6.140}$$

Let us now calculate how the expectation value of the momentum operator changes with the evolution parameter τ. The procedure is analogous to that in Sec. 6.4. Instead of the state (6.83) we now take

$$|\psi\rangle = \prod_\xi \int d^D p_{(\xi)} \, g^{(\xi)}(p_{(\xi)}) a^\dagger_{(\xi)}(p_{(\xi)})|0\rangle, \qquad \text{no integration over } (\xi). \tag{6.141}$$

Taking $m = m(\xi)$ and introducing

$$h(p, \xi) = \frac{\Lambda}{2}(p^2 - m^2), \tag{6.142}$$

we obtain

$$\frac{d}{d\tau}\langle\psi|\hat{p}_\mu|\psi\rangle = (-i) \int d^D p \, p_\mu \, g^*(\xi, p) g(\xi', p) s(\xi, \xi')(h(p, \xi) - h(p, \xi')) d\xi d\xi', \tag{6.143}$$

where $g(\xi, p) \equiv g^{(\xi)}(p)$ and $s(\xi, \xi') \equiv s_{(\xi)(\xi')}$.

If we take the field space metric

$$s(\xi, \xi') = (1 + \lambda_c \partial^a \partial_a) \, \delta^d(\xi - \xi'), \tag{6.144}$$

then

$$\frac{d}{d\tau}\langle p_\mu\rangle \equiv \frac{d}{d\tau}\langle\psi|\hat{p}_\mu|\psi\rangle = (-i)\lambda_c \int d^D p \, d^p \xi \, p_\mu \, h(p, \xi)(g^* \partial_a \partial^a g - \partial_a \partial^a g^* g). \tag{6.145}$$

This is the time derivative of the expectation value of the total momentum of the brane, and it vanishes if $h(p, \xi)$ does not change with ξ.

The expectation value of the total momentum is given by the integral over the momenta of the brane's segments:

$$\langle\psi|\hat{p}_\mu|\psi\rangle = \int dp \, d\xi \, d\sigma' \, p_\mu \, g^*(\xi, p) g(\xi', p) s(\xi, \xi')$$

$$= \int dp \, d\xi \, p_\mu (g^* g + g^* \partial_a \partial^a g) = \langle \hat{p}_\mu\rangle = \int d\xi \langle \hat{p}_\mu\rangle_\xi \tag{6.146}$$

where

$$\langle \hat{p}_\mu \rangle_\xi = \int \mathrm{d}p\, p_\mu (g^* g + g^* \partial_a \partial^a g) = \langle \psi |_\xi \hat{p}_\mu | \psi \rangle_\xi. \tag{6.147}$$

From Eq. (6.145) we then read the following expression for the time derivative of the momentum of a brane segment:

$$\frac{\mathrm{d}}{\mathrm{d}\tau} \langle p_\mu \rangle_\xi = (-i) \lambda_c \int \mathrm{d}^D p\, p_\mu\, h(p, \xi)(g^* \partial_a \partial^a g - \partial_a \partial^a g^* g), \tag{6.148}$$

which in general is different from zero even if $h(p, \xi)$ does not change with ξ.

If in Eq. (6.148) we express the wave packet profile $g(\xi, p)$ in term of its position space counter part $f(\xi, p)$,

$$g(\xi, p) = \frac{1}{(2\pi)^{D/2}} \int e^{-i p_\mu x^\mu} f(\xi, x) \mathrm{d}x, \tag{6.149}$$

then we obtain

$$\frac{\mathrm{d}}{\mathrm{d}\tau} \langle \hat{p}_\mu \rangle_\xi = -\lambda_c \partial_a \int \mathrm{d}^D x \left[f^*(\xi, x) \left(\frac{1}{2\Lambda} (\partial_\mu \partial^\mu - m^2) \partial^a \partial_\mu f \right) \right.$$
$$\left. - \left(\frac{1}{2\Lambda} (\partial_\mu \partial^\mu - m^2) \partial^a \partial_\mu f^* \right) f \right]. \tag{6.150}$$

Though not written explicitly, the wave packet profiles g and f depend on the evolution time τ. Using the Schrödinger equation with the Hamiltonina (6.139) for the state (6.141), we obtain the equation of motion for the wave packet profile f:

$$\frac{1}{2\Lambda} (\partial_\mu p^\mu - m^2) f = -i \frac{\partial}{\partial \tau} f. \tag{6.151}$$

Using the latter equation in Eq. (6.150) we obtain

$$\frac{\mathrm{d}}{\mathrm{d}\tau} \langle \hat{p}_\mu \rangle_\xi = -\lambda_c \partial_a \int \mathrm{d}^D x \left[f^*(\xi, x) \left(-i \frac{\partial}{\partial \tau} \partial^a \partial_\mu f \right) - \left(-i \frac{\partial}{\partial \tau} \partial^a \partial_\mu f^* \right) f \right]. \tag{6.152}$$

For a Gaussian wave packet profile

$$f \approx A e^{-\frac{(x - \bar{X}(\xi))^2}{2\sigma_0}} e^{i \bar{p}_\mu x^\mu} e^{i \bar{h}, \tau}, \tag{6.153}$$

where

$$\bar{h} = \frac{1}{2\Lambda} (\bar{p}^2 - m^2), \tag{6.154}$$

equation (6.152) becomes

$$\frac{\langle \hat{p}_\mu \rangle_\xi}{\mathrm{d}\tau} = -\lambda_c \partial_a \left(\frac{\bar{h}}{\Delta S} \frac{\partial^a \bar{X}_\mu}{\tilde{\sigma}_0} \right). \tag{6.155}$$

The expressions with the metric (6.144) are not covariant with respect to arbitrary reparametrizations of ξ^a. If we take the metric

$$s(\xi, \xi') = \sqrt{-\gamma(\xi)}\delta^d(\xi - \xi') + \partial_a \left(\sqrt{-\gamma(\xi)}\gamma^{ab}\partial_b \right) \delta^d(\xi - \xi') \qquad (6.156)$$

where $\gamma \equiv \det\gamma_{ab}$, then the expressions become covariant, and instead of (6.155) we obtain

$$\frac{\langle \hat{p}_\mu \rangle_\xi}{d\tau} = -\lambda_c \partial_a \left(\frac{\bar{h}}{\Delta S} \frac{\sqrt{-\gamma(\xi)}\gamma^{ab}\partial_b \bar{X}_\mu}{\tilde{\sigma}_0} \right). \qquad (6.157)$$

The latter equation tells how the expected momentum density $\langle \hat{p}_\mu \rangle_\xi$ changes with the evolution parameter τ, which in the Stueckelberg theory is the "true" time, whereas $x^0 \equiv t$ is just one of spacetime coordinates. In Appendix we show that Eq. (6.157) corresponds to the equation of motion of a classical Stueckelberg brane (see [49]), which is a generalization of the Stueckelberg point particle.

6.5.3 *Self interacting Stueckelberg field in configuration space*

In the absence of interactions, a field $\psi(\tau, x^M) \equiv \psi(\tau, x^{1\mu}, x^{2\mu}, ..., x^{N\mu})$ over a many particle configuration is the product (6.121) of the single particle fields. In the presence of interactions, in general this is no longer the case. An interacting field theory is described by the action (6.118) to which we add an interactive term I_{int}, so that the total action is

$$I = I_0 + I_{\text{int}}. \qquad (6.158)$$

We will take $I_{\text{int}} = -\frac{G_0}{4!}(\psi^*\psi)^2$. Let us also assume that a particle, say, No. 1, can be singled out from the rest of the configuration according to (6.123). Inserting Eq. (6.123) into the action (6.158), we obtain the Stueckelberg action for the scalar field $\varphi(\tau, x^\mu)$ with the quartic self interaction:

$$I = \int d\tau\, d^4x \left[-i2\Lambda\varphi^*\frac{\partial\varphi}{\partial\tau} + \partial_\mu\varphi^*\partial^\mu\varphi + m_{\text{res}}\varphi^*\varphi - g_0(\varphi^*\varphi)^2 \right], \qquad (6.159)$$

where $g_0 = G_0 \int d^{D-4}\bar{x}(\chi^*\chi)^2$, and

$$m_{\text{res}}^2 = \int d^{D-4}\bar{x} \left(\partial_{\bar{M}}\chi^*\partial^{\bar{M}}\chi - i2\Lambda\chi^*\frac{\partial\chi}{\partial\tau} \right), \qquad (6.160)$$

is the residual mass that is determined by the presence of the field $\chi(\tau, x^{\bar{M}})$ due to all the other particles of the configuration. In general, m_{res}^2 is different from zero. In particular, in the absence of an interaction, χ is given by Eq. (6.124) and then $m_{\text{res}}^2 = 0$.

For an interacting field theory the factorization (6.121) of a field $\psi(\tau, x^M)$ is valid only if the particle No. 1 is not entangled with the other, mutually interacting, particles. If it is entangled, then (6.121) does not hold. We must then work with the field $\psi(\tau, x^M)$ without factoring out a single particle field.

We have thus arrived at the many particle analog of the brane theory, described by the classical action (6.22) or the first quantized action (6.48), in which now the metric $\rho_{\mu(\sigma)\nu(\sigma')}$ of the brane space is not flat. Then one cannot describe a brane as a bunch of point particles. Similarly, in general one cannot describe a many particle configuration as a bunch of point particles. Only if the metric is $g_{MN} \equiv g_{(i\mu)(j\nu)} = \delta_{ij}g_{\mu\nu}$ one has a bunch of point particles. In general, the metric need not be diagonal in the indices $(i\mu)$, $(j\nu)$. Then the particles are intertwined more than it is usually assumed. The physics, either classical or quantized, has to be done in a configuration space $\mathcal{C}$ of many particles/branes. The metric ρ_{MN} of $\mathcal{C}$ in general is curved. An interactive term such as $I_{\text{int}} = -\frac{G_0}{4!}(\psi^*\psi)^2$ can be obtained from the dimensional reduction of the action of the form (6.118), along the lines similar to that of Ref. [11].

6.6   Conclusion

Within this approach configuration space $\mathcal{C}$ is primary even in classical physics, and the action principle must be formulated in $\mathcal{C}$, not in spacetime. In other words, physics, both classical and quantum, must be formulated in configuration space which can be a space of many point particles and/or branes. Space or spacetime is a subspace of a configuration space (Fig. 6.9). The concept of spacetime has to be revised by considering spacetime as a subspace of a configuration space, which ultimately is that of the whole universe.

We have arrived at such conclusion by inspecting the action of a Dirac-Nambu-Goto brane. We have found that a brane can be considered as a point in an infinite dimensional brane space $\mathcal{M}$, moving along a geodesic in $\mathcal{M}$. The metric of $\mathcal{M}$ is not fixed, it is dynamical, like in general relativity. For a particular metric we obtain the usual Dirac-Nambu-Goto brane. More general metrics give us interesting fancy branes (Fig. 6.3) that might be useful in scenarios for quantum gravity in the presence of matter, where matter is given by the brane's self intersections [49, 153]. The simplest is the "flat" metric that gives us "flat branes" (Fig. 6.2). A flat brane can be straightforwardly quantized as a bunch of point particles. If we take suitable interactions between the quantum fields, we obtain as an "expectation value" the classical Dirac-Nambu-Goto brane.

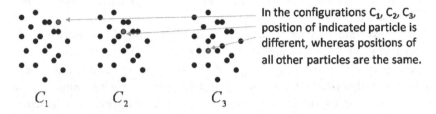

In the configurations C_1, C_2, C_3, position of indicated particle is different, whereas positions of all other particles are the same.

C_1 C_2 C_3

Fig. 6.9   Space(time) as a subspace of configuration space. Configuration space is the space of all possible configurations, whereas (space)time is the space of only those possible configurations in which the positions of all except one particle are considered as fixed.

We have thus found how to quantize branes: via flat brane space. Non flat branes are then objects of an effective classical theory that arises from the underlying QFT of many interacting fields.

The concept of configuration space is associated not only with branes, but with whatever physical systems, in classical and quantum theory. A configuration can be:

- a single brane, considered as a bunch of point particles,
- a discrete system of point particles,
- a mixed system of many branes and point particles,
- etc.

A closed brane or a system of closed branes (Fig. 6.8) can be approximately described by a finite number of degrees of freedom, which are coordinates of the 16-dimensional Clifford space. The latter space has signature (8,8), i.e., its points can be described by eight "time like" coordinates (associated with the plus sign of the metric) and eight "space like" coordinates (associated with the minus sign of the metric). By picking up one time like and one space like coordinate, and composing from them the analog of two light-cone coordinates, we have derived the Stueckelberg action for a scalar field. We have also shown how a continuous set of such locally interacting fields leads to the effective classical branes à la Stueckelberg. The latter objects satisfy the equations of motion that can be obtained by calculating the time derivative of the expectation value of the momentum operator with respect to certain "wave packet" like quantum states created by the Stueckelberg field operators.

Appendix: Stueckelberg point particle and its generalization to a brane

The phase space action for a point particle in a $(D+2)$-dimensional space with signature $(2, D)$ is

$$I = \int d\tau \left[p_M \dot{x}^M - \frac{\alpha}{2}(p_M p^M - M^2) \right], \tag{6.161}$$

where α is a Lagrange multiplier whose variation gives the constraint $p_M p^M - M^2 = 0$. The signature of the extra two dimensions is $(+-)$, whilst the signature of the D-dimensional space is $(1, D-1)$. Now we take $D = 4$, so that we have an extra fifth and sixth dimension. If x^5 and x^6 are "light-cone" coordinates, then the action reads

$$I = \int d\tau \left[p_\mu \dot{x}^\mu + p_5 \dot{x}^5 + p_6 \dot{x}^6 - \frac{\alpha}{2}(p_\mu p^\mu - 2p_5 p_6 - M^2) \right]. \tag{6.162}$$

Taking a gauge in which $\tau = x^5$, we have

$$p_6 \equiv -\Lambda = \frac{\dot{x}_6}{\alpha}; \quad x^5 = \tau; \quad \alpha = -\frac{\dot{x}_6}{\Lambda} = \frac{\dot{x}^5}{\Lambda} = \frac{1}{\Lambda}. \tag{6.163}$$

The action (6.162) can then be written as

$$I = \int d\tau \left[p_\mu \dot{x}^\mu + p_5 - \frac{1}{2\Lambda}(p_\mu p^\mu - M^2) \right]. \tag{6.164}$$

Here Λ is not a Lagrange multiplier, but a fixed quantity, namely $\Lambda \equiv -p_6$.

But we can omit p_5 in the above action, because it does not contribute to the x^μ equations of motion. Then we have

$$I = \int d\tau \left[p_\mu \dot{x}^\mu - \frac{1}{2\Lambda}(p_\mu p^\mu - M^2) \right]. \tag{6.165}$$

The corresponding Hamiltonian is

$$H = \frac{1}{2\Lambda}(p_\mu p^\mu - M^2). \tag{6.166}$$

The above action is the Stueckelberg action. It is derived from the higher dimensional action.

From the constraint $p_M p^M - M^2 = p_\mu p^\mu - 2p_5 p_6 - M^2$ we have

$$p_5 = \frac{1}{2p_6}(p_\mu p^\mu - M^2) = -H, \tag{6.167}$$

which means that the Hamiltonian is given by the fifth component of momentum, and is thus a generator of translations along $x^5 = \tau$, whilst the constant Λ is given by the sixth component of momentum.

Now let us do the same for a brane. Let $\xi \equiv \xi^a$, $a = 1, 2, ..., d$, be d parameters of a brane in $(D + 2)$-dimensions. Now a brane need not be space like. It can extend either into space like or into time like directions, or both [49]. If $D = 4$ then the extra two dimensions are x^5 and x^6, but we may keep the same notation for the extra two dimensions even if $D > 4$. We then have $x^M = (x^\mu, x^5, x^6)$, $\mu = 0, 1, 2, 3, 7, 8, ..., D - 3$.

The phase space brane action is

$$I = \int \mathrm{d}\tau \mathrm{d}^d\xi \left[p_M \dot{x}^M - \frac{\alpha}{2\kappa\sqrt{-\gamma}}(p_M p^M - \kappa^2(-\gamma)) \right]$$

$$= \int \mathrm{d}\tau \mathrm{d}^d\xi [p_\mu \dot{x}^\mu + p_5 \dot{x}^5 + p_6 \dot{x}^6$$

$$- \frac{\alpha}{2\kappa\sqrt{-\gamma}}(p_\mu p^\mu - 2p_5 p_6 - \kappa^2(-\gamma)]. \tag{6.168}$$

Choosing a gauge $\tau = x^5$, and using

$$p_6 = -\sqrt{-\gamma}\tilde{\Lambda}; \quad \dot{x}_M = \frac{\alpha}{\kappa\sqrt{-\gamma}}p_M \; ; \dot{x}_6 = \frac{\alpha}{\kappa\sqrt{-\gamma}}(-\sqrt{-\gamma}\tilde{\Lambda}) = -\dot{x}^5 = -1, \tag{6.169}$$

we obtain

$$I = \int \mathrm{d}\tau \mathrm{d}^d\xi \left[p_\mu \dot{x}^\mu - p_5 - \frac{1}{2\sqrt{-\gamma}\tilde{\Lambda}}(p_\mu p^\mu - \kappa^2(-\gamma)) \right]. \tag{6.170}$$

Let us omit p_5, because this term does no influence the x^μ equations of motion. Then we obtain the following unconstrained (Stueckelberg) action,

$$I = \int \mathrm{d}\tau \mathrm{d}^d\xi \left[p_\mu \dot{x}^\mu - \frac{1}{2\sqrt{-\gamma}\tilde{\Lambda}}(p_\mu p^\mu - \kappa^2(-\gamma)) \right] \tag{6.171}$$

which is a generalized of the Stueckelberg point particle action. The corresponding Hamiltonian is

$$H = \int \mathrm{d}^d\xi (p_\mu \dot{x}^\mu - L) = \int \mathrm{d}^d\xi \frac{1}{2\sqrt{-\gamma}\tilde{\Lambda}}(p_\mu p^\mu - \kappa^2(-\gamma)) = \int \mathrm{d}^d\xi \, \mathcal{H}. \tag{6.172}$$

From the constraint $p_\mu p^\mu - 2p_5 p_6 - \kappa^2(-\gamma) = 0$ we have

$$p_5 = \frac{1}{2p_6}(p_\mu p^\mu - \kappa^2(-\gamma)) = -\mathcal{H}. \tag{6.173}$$

The Hamiltonian of a brane segment is

$$h = \Delta\xi \frac{1}{2\sqrt{-\gamma}\tilde{\Lambda}}(p_\mu(\xi)p^\mu(\xi) - \kappa^2(-\gamma)). \tag{6.174}$$

Here $p_\mu(\xi)$ is the momentum density. We introduce the momentum and the mass of a brane segment

$$p_\mu = p_\mu(\xi)\Delta\xi \ , \qquad m = \kappa\sqrt{-\gamma}\Delta\xi. \tag{6.175}$$

We also define

$$\frac{p_\mu(\xi)}{\tilde{\Lambda}} = \frac{p_\mu}{\Lambda} = \frac{p_\mu(\xi)\Delta\xi}{\Lambda} \tag{6.176}$$

from which it follows

$$\Lambda = \tilde{\Lambda}\Delta\xi. \tag{6.177}$$

The Hamiltonian of a brane segment thus becomes

$$h = \frac{1}{2\Lambda}(p_\mu p^\mu - m^2). \tag{6.178}$$

Equation of motion derived from the Stueckelberg brane action (6.171) is

$$\frac{\mathrm{d}p_\mu(\xi)}{\mathrm{d}\tau} + \partial^a \left[\frac{1}{2\tilde{\Lambda}} \left(\frac{1}{\sqrt{-\gamma}} p^\mu p_\mu - \kappa^2 \sqrt{-\gamma} \right) \partial^a x_\mu \right] = 0. \tag{6.179}$$

This corresponds to the quantum expectation value equation (6.157).

Chapter 7

Particle Position in Quantum Field Theories

In the previous two chapters we have seen that branes can be created by the action of field operators on a vacuum. In such a way one can, for instance, create a 3-brane sweeping a 4-dimensional surface that, according to the braneworld scenario, represents our spacetime. We have also seen that matter in such setup arises as holes within the brane. Successful quantization of a generic brane would be a major step towards quantum gravity. Unfortunately, this has turned out to be a very tough problem that has not yet been completely resolved so far. But, as we have seen, by using the concept of brane created by an infinite set of field operators, considered as creation operators that create particles at certain positions, quantizing a brane reduces to the problem of interacting quantum field theory. But according to the usual understanding, operators that create particles at definite positions make no sense in relativistic quantum field theories. We will now show how in Refs. [166, 167] the issue concerning position in QFT has been carefully examined and shown that the conventional arguments have to be revised.

7.1 Manifestly covariant formulation of a scalar field quantum theory

Quantum field theory of a Hermitian scalar field operator $\varphi(x)$ is described by the action

$$I = \frac{1}{2} \int d^4x \left(\partial_\mu \varphi \partial^\mu - m^2 \varphi^2 \right), \tag{7.1}$$

where x denotes a spacetime point and d^4x a spacetime volume element. The corresponding equation of motion is the Klein-Gordon equation,

$$\left(\partial_\mu \partial^\mu + m^2 \right) \varphi = 0, \tag{7.2}$$

141

whose solution is

$$\varphi(x) = \int \frac{\mathrm{d}^4 p}{(2\pi)^4} c(p)\delta(p^2 - m^2)e^{-ip_\mu x^\mu}. \qquad (7.3)$$

Usually a spacetime point is labelled by four coordinates $x \equiv x^\mu$, $\mu = 0, 1, 2, 3$. The spacetime and the momentum space 4-volumes are then written as $\mathrm{d}^4 x = \mathrm{d}t\,\mathrm{d}^3\boldsymbol{x}$ and $\mathrm{d}^4 p = \mathrm{d}p^0\,\mathrm{d}^3\boldsymbol{p}$. This means that we have split spacetime into time, parametrized by $x^0 \equiv t$, and space, parametrized by x^i, $i = 1, 2, 3$. More generally, we can split spacetime independently of the coordinates, associated with a given Lorentz reference frame, by introducing a unit time like 4-vector n^μ, normal to a space like 3-surface Σ. The coordinates and momenta can then be projected into the part orthogonal to Σ,

$$x_\perp \equiv s = x^\mu n_\mu \,, \qquad\qquad p_\perp = p_\mu n^\mu, \qquad (7.4)$$

and the part, tangential to Σ:

$$\bar{x}^\mu = P^\mu{}_\nu x^\nu \,, \qquad\qquad \bar{p}^\mu = P_\mu{}^\nu p_\nu. \qquad (7.5)$$

Here $P^\mu{}_\nu = \delta^\mu{}_\nu - n^\mu n_\nu$ is the projector onto Σ. Thus

$$x = (s, \bar{x}^\mu) \,, \qquad\qquad p = (p_\perp, \bar{p}_\mu). \qquad (7.6)$$

Similarly we can split the partial derivative:

$$\partial = (\partial_\perp, \bar{\partial}_\mu) \,, \qquad \partial_\perp \equiv \frac{\partial}{\partial s} = n^\mu \partial_\mu \,, \qquad \bar{\partial} = P_\mu{}^\nu \partial_\nu. \qquad (7.7)$$

In particular, if $n^\mu = (1, 0, 0, 0)$, then $x = (x^0, x^i) \equiv (t, \boldsymbol{x})$, $\partial = \left(\frac{\partial}{\partial t}, \frac{\partial}{\partial x^i}\right)$, $x^0 \equiv t$, $x^i \equiv \boldsymbol{x}$, $i, j = 1, 2, 3$.

In general, n^μ is an arbitrary unit time like vector. The volume element can then be factorized as

$$\mathrm{d}^4 x = \mathrm{d}s\,\mathrm{d}\Sigma \,, \qquad\qquad \mathrm{d}^4 p = \mathrm{d}p_\perp \mathrm{d}\Sigma_{\bar{p}}, \qquad (7.8)$$

where

$$\mathrm{d}\Sigma = n^\mu \epsilon_{\mu\nu\rho\sigma} \mathrm{d}\bar{x}^\nu \mathrm{d}\bar{x}^\rho \mathrm{d}\bar{x}^\sigma \quad \text{and} \quad \mathrm{d}\Sigma_{\bar{p}} = n_\mu \epsilon^{\mu\nu\rho\sigma} \mathrm{d}\bar{p}_\nu \mathrm{d}\bar{p}_\rho \mathrm{d}\bar{p}_\sigma \qquad (7.9)$$

are the volume elements within the 3-surface that is orthogonal to n_μ.

Using (7.7), the action (7.1) can be written as

$$I = \frac{1}{2} \int \mathrm{d}\Sigma \left(\left(\frac{\partial\varphi}{\partial s}\right)^2 + \bar{\partial}_\mu \varphi \bar{\partial}^\mu \varphi - m^2 \varphi^2 \right)$$

$$= \frac{1}{2} \int \mathrm{d}\Sigma \left(\left(\frac{\partial\varphi}{\partial s}\right)^2 - \varphi \left(m^2 + \bar{\partial}_\mu \bar{\partial}^\mu\right)\varphi \right) \qquad (7.10)$$

In the last step a partial integration was performed and the surface term omitted.

The canonically conjugated variables are φ and $\Pi = \partial/\partial s$. They satisfy the following commutation relations at equal s:

$$[\varphi(s,\bar{x}), \Pi(s,\bar{x}')] = i\delta^4(\bar{x} - \bar{x}'), \tag{7.11}$$

$$[\varphi(s,\bar{x}), \varphi(s,\bar{x}')] = 0 , \qquad [\Pi(s,\bar{x}), \Pi(s,\bar{x}')] = 0. \tag{7.12}$$

The Hamilton operator associated with the action (7.10) is

$$H = \frac{1}{2} \int d\Sigma \left(\Pi^2 + \varphi \left(m^2 + \bar{\partial}_\mu \bar{\partial}^\mu \right) \varphi \right). \tag{7.13}$$

The quantum theory of the scalar field as formulated by the action (7.10) and the corresponding Hamilton operator (7.13) is covariant, because it has the same form in any Lorentz frame, i.e., it does not change if we change the coordinates by a Lorentz transformation $x^\mu \to x'^\mu = L^\mu{}_\nu x^\nu$.

The system (7.10), (7.13) describes an infinite uncountable set of harmonic oscillators, distinguishe by the label $\bar{x} \equiv \bar{x}^\mu$. Let us therefore introduce the operators[1]

$$a(s,\bar{x}) = \frac{1}{\sqrt{2}} \left(\sqrt{\omega_{\bar{x}}}\varphi + \frac{i}{\sqrt{\omega_{\bar{x}}}}\Pi \right), \tag{7.14}$$

$$a^\dagger(s,\bar{x}) = \frac{1}{\sqrt{2}} \left(\sqrt{\omega_{\bar{x}}}\varphi - \frac{i}{\sqrt{\omega_{\bar{x}}}}\Pi \right), \tag{7.15}$$

where

$$\omega_{\bar{x}} = \sqrt{m^2 + \bar{\partial}_\mu \bar{\partial}^\mu}. \tag{7.16}$$

They satisfy

$$[a(s,\bar{x}), a^\dagger(s,\bar{x}')] = \delta^4(\bar{x} - \bar{x}'), \tag{7.17}$$

$$[a(s,\bar{x}), a(s,\bar{x}')] = 0 , \qquad [a^\dagger(s,\bar{x}), a^\dagger(s,\bar{x}')] = 0. \tag{7.18}$$

Expressed in terms of those operators the Hamilton operator (7.13) becomes

$$H = \frac{1}{2} \int d\Sigma \left(a^\dagger(s,\bar{x})\omega_{\bar{x}}a(s,\bar{x}) + a(s,\bar{x})\omega_{\bar{x}}a^\dagger(s,\bar{x}) \right). \tag{7.19}$$

---

[1]In particular, if $n^\mu = (1,0,0,0)$, then $s = x^0 = t$, $\bar{x} = (0, x^i) \equiv (0, \boldsymbol{x})$, then the Hamiltonian (7.13) and the operators given below become those introduced by Jackiw [168].

Instead of position space representation, one can as well use momentum space representation, and transform the operators according to

$$a(s, \bar{x}) = \frac{1}{\sqrt{(2\pi)^3}} \int d\Sigma_p a(\bar{p}) e^{ipx}, \tag{7.20}$$

$$a^\dagger(s, \bar{x}) = \frac{1}{\sqrt{(2\pi)^3}} \int d\Sigma_p a^\dagger(\bar{p}) e^{-ipx}, \tag{7.21}$$

where $px = p_\mu x^\mu = p_\perp s + \bar{p}_\mu$, $x_\perp \equiv s$. By inserting (7.20), (7.21) into the Hamilton operator (7.19), we obtain

$$H = \frac{1}{2} \int d\Sigma_{\bar{p}}\, \omega_{\bar{p}} \left(a^\dagger(\bar{p}) a(\bar{p}) + a(\bar{p}) a^\dagger(\bar{p}) \right). \tag{7.22}$$

Using (7.14), (7.15), and (7.20), (7.21), we have

$$\varphi(s, \bar{x}) = \frac{1}{\sqrt{2\omega_{\bar{x}}}} \left(a(s, \bar{x}) + a^\dagger(s, \bar{x}) \right)$$

$$= \int d\Sigma_{\bar{p}} \frac{1}{\sqrt{(2\pi)^3 2\omega_{\bar{p}}}} \left(a(\bar{p})^{-ipx} + a^\dagger(\bar{p})^{-ipx} \right). \tag{7.23}$$

The same expression for the scalar field $\varphi(s, \bar{x})$ in terms of the momentum space operators also comes directly from Eq. (7.3).

If we rewrite Eq. (7.23) in terms of the operators

$$\tilde{a}(\bar{p}) = \sqrt{(2\pi)^3 2\omega_{\bar{p}}} a(\bar{p}), \qquad \tilde{a}^\dagger(\bar{p}) = \sqrt{(2\pi)^3 2\omega_{\bar{p}}} a^\dagger(\bar{p}), \tag{7.24}$$

we obtain

$$\varphi(s, \bar{x}) = \int d\Sigma_{\bar{p}} \frac{1}{(2\pi)^3 2\omega_{\bar{p}}} \left(\tilde{a}(\bar{p})^{-ipx} + \tilde{a}^\dagger(\bar{p})^{-ipx} \right), \tag{7.25}$$

which, in the case when $n^\mu = (1, 0, 0, 0)$, becomes the usual, most often used, field expansion.

Between the operators $a(\bar{p})$, $a^\dagger(\bar{p})$ and $a(\bar{x})$, $a^\dagger(\bar{x})$ there are the covariant relations (7.20), (7.21). In such a covariant formulation of the theory it is obvious that both representations, namely, the coordinate one in which H is given by (7.13) or (7.19), and the momentum representation in which we have (7.22), are equally legitimate. Therefore, also the operators $a(\bar{x})$, $a^\dagger(\bar{x})$ are as legitimate as the operators $a(\bar{p})$, $a^\dagger(\bar{p})$. In particular Lorentz frame in which $n^\mu = (1, 0, 0, 0)$, we have $\bar{x}^\mu = (0, \boldsymbol{x})$, $\bar{p}_\mu = (0, \boldsymbol{p})$, and the operators are $a(\boldsymbol{x})$, $a^\dagger(\boldsymbol{x})$, $a(\boldsymbol{p})$, $a^\dagger(\boldsymbol{p})$. According to the usual understanding of quantum field theory, the momentum space annihilation/creation operators are acceptable ingredients of the theory, whilst the position space annihilation/creation operators and the associated position operator and position state make no sense. However, within the above manifestly covariant formulation of quantum field theory it is clear that also the operators $a(\bar{x})$, $a^{\dagger}(\bar{x})$, in particular, $a(\boldsymbol{x})$, $a^{\dagger}(\boldsymbol{x})$, are unavoidable ingredients, and cannot be swept under the carpet.

Whilst the operators $a(\bar{x})$ annihilate the vacuum, defined according to $a(\bar{x})|0\rangle = 0$, the operators $a^{\dagger}(\bar{x})$ create the states with definite position $\bar{x} \equiv \bar{x}^{\mu}$ on a hypersurface Σ:

$$|\bar{x}\rangle = a^{\dagger}(\bar{x})|0\rangle. \tag{7.26}$$

The corresponding relation in momentum space is

$$|\bar{p}\rangle = a^{\dagger}(\bar{p})|0\rangle, \tag{7.27}$$

denotes the states with the momentum, whose space like component, $\bar{p}^{\mu}$, is within the hypersurface Σ.

The states (7.26) are eigenstates of the operator

$$\hat{\bar{x}}^{\mu} = \int d\Sigma \, a^{\dagger}(\bar{x}) \bar{x}^{\mu} a(\bar{x}), \tag{7.28}$$

which, in particular if $n^{\mu} = (1,0,0,0)$ and thus $\bar{x} = (0, \boldsymbol{x})$, assumes the usual form of the Newton-Wigner position operator [169]

$$\hat{\boldsymbol{x}} = \int d\Sigma \, a^{\dagger}(\boldsymbol{x}) \boldsymbol{x} a(\boldsymbol{x}), \tag{7.29}$$

whose eigenstates are $|\boldsymbol{x}\rangle = a^{\dagger}(\boldsymbol{x})|0\rangle$.

Position states $|\bar{x}\rangle$ are defined with respect to a hypersurface Σ. So are also defined momentum states $|\bar{p}\rangle$. A hypersurface Σ is the simultaneity 3-surface of a given observer $\mathcal{O}_n$, whose proper time direction is along the normal n^{μ} to Σ. The vector $\bar{x}^{\mu}$ denotes positions on Σ, and all those positions are mutually simultaneous relative to $\mathcal{O}_n$. They form the 3-space experienced by $\mathcal{O}_n$. In Ref. [167] it stays:

> That spatial positions refer to a given 3-surface (hypersurface) Σ in spacetime, is incorporated in the formulation of relativity. We talk about the observers, simultaneity surface, length contraction, etc., where "length" refers to a distance within a given Σ, and "contractions" refers to how such a length looks in another frame.

Equation (7.26) is a covariant description of position states, valid in the Lorentz reference frame in which position on a given 3-surface Σ are determined by the 4-vector $\bar{x}^{\mu}$. If observed from another Lorentz frame, the same positions are determined by a different set of coordinates, namely

$\bar{x}'^{\mu}$, related to $\bar{x}^{\mu}$ by a Lorentz transformation $\bar{x}'^{\mu} = L^{\mu}{}_{\nu}\bar{x}^{\nu}$ (shortly $\bar{x} = L\bar{x}$). The position states on Σ are thus described in a new Lorentz frame according to

$$|\bar{x}\rangle = a'(\bar{x}')|0\rangle, \tag{7.30}$$

where

$$a'^{\dagger}(\bar{x}') = a^{\dagger}(\bar{x}) = a^{\dagger}(L^{-1}\bar{x}'). \tag{7.31}$$

Successive actions of the operators $a^{\dagger}(\bar{x})$ on the vacuum $|0\rangle$ create a multiparticle state

$$|\bar{x}_1, \bar{x}_2, \bar{x}_3, ...\rangle = a^{\dagger}(\bar{x}_1)a^{\dagger}(\bar{x}_2)a^{\dagger}(\bar{x}_3)...|0\rangle, \tag{7.32}$$

with definite positions on Σ. Equivalently, successive actions of the operators $a^{\dagger}(\bar{p})$ on the vacuum create a multiparticle state $|\bar{p}_1, \bar{p}_2, \bar{p}_3, ...\rangle$ with definite momenta tangential to Σ.

7.2 States with indefinite position or momentum

7.2.1 *Wave packet profiles for a Hermitian and non Hermitian scalar field*

Physical states cannot have exact positions or momenta. In general they are superpositions. In the case of a *Hermitian scalar field* we have

$$|\Psi\rangle = \sum_{r=1}^{N} d\Sigma_1 d\Sigma_2...d\Sigma_r f(s, \bar{x}_1, \bar{x}_2, ..., \bar{x}_r)a^{\dagger}(\bar{x}_1)a^{\dagger}(\bar{x}_2)...a^{\dagger}(\bar{x}_r)|0\rangle$$

$$= \sum_{r=0}^{N} \int d\Sigma_{\bar{p}_1} d\Sigma_{\bar{p}_2}...d\Sigma_{\bar{p}_r} g(s, \bar{p}_1, \bar{p}_2, ..., \bar{p}_r)a^{\dagger}(\bar{p}_1)a^{\dagger})\bar{p}_2)...a^{\dagger}(\bar{p}_r)|0\rangle, \tag{7.33}$$

where $f(s, \bar{x}_1, \bar{x}_2, ..., \bar{x}_r)$ and $g(s, \bar{p}_1, \bar{p}_2, ..., \bar{p}_r)$ are complex valued wave packet profiles (wave functions) in position and momentum space, respectively.

For a single particle state we have

$$|\Psi\rangle = \int d\Sigma\, f(s, \bar{x})a^{\dagger}(\bar{x})|0\rangle = \int d\Sigma_{\bar{p}}\, g(s, \bar{p})a^{\dagger}(\bar{p})|0\rangle. \tag{7.34}$$

The wave packet profiles f and g can be Fourier transformed into each other according to

$$f(s, \bar{x}) = \frac{1}{\sqrt{(2\pi)^3}} \int d\Sigma_{\bar{p}}\, g(s, \bar{p})e^{i\bar{p}_{\mu}\bar{x}^{\mu}}. \tag{7.35}$$

The state (7.34) can as well be expressed in terms of the operators $\tilde{a}^\dagger(\bar{p})$, related to $a^\dagger(\bar{p})$ according to Eq. (7.25), the corresponding relation for the wave packet profile being

$$\tilde{g}(s,\bar{p}) = \sqrt{(2\pi)^3 2\omega_{\bar{p}}}\, g(s,\bar{p}). \tag{7.36}$$

Equivalently, the state (7.34) can be expressed in terms of the position space operators and wave packet profiles, renormalized according to

$$\tilde{a}^\dagger(\bar{x}) = (2\omega_{\bar{x}})^{-1/2} a^\dagger(\bar{x})\,, \qquad \tilde{f}(s,\bar{x}) = (2\omega_{\bar{x}})^{-1/2} f(s,\bar{x}). \tag{7.37}$$

Thus, Eq. (7.34) can also be written as

$$|\Psi\rangle = \int \frac{\mathrm{d}\Sigma_{\bar{p}}}{(2\pi)^3 2\omega_{\bar{p}}}\, \tilde{g}(s,\bar{p})\tilde{a}^\dagger(\bar{p})|0\rangle = \int \mathrm{d}\Sigma \sqrt{2\omega_{\bar{x}}}\, \tilde{f}(s,\bar{x}) \sqrt{2\omega_{\bar{x}}}\, \tilde{a}^\dagger(\bar{x})|0\rangle. \tag{7.38}$$

A generic state $|\Psi\rangle$ satisfies the Schrödinger equation

$$i\frac{|\Psi\rangle}{\partial s} = H|\Psi\rangle, \tag{7.39}$$

where H is the Hamilton operator, defined in Eq. (7.13), that can also be expressed as in Eqs. (7.19) or (7.22).

If we insert the expression for a generic multiparticle state (7.33) into the Schrödinger equation (7.39) and omit the zero point energy, we obtain a set of equations for r-particle states:

$$i\frac{\partial f(s,\bar{x}_1,\bar{x}_2,...,\bar{x}_r)}{\partial s} = \sum_{k=1}^{r} \omega_{\bar{x}_k} f(s,\bar{x}_1,\bar{x}_2,...,\bar{x}_k,...\bar{x}_r), \tag{7.40}$$

$$i\frac{\partial g(s,\bar{p}_1,\bar{p}_2,...,\bar{p}_r)}{\partial s} = \sum_{k=1}^{r} \omega_{\bar{p}_k} g(s,\bar{p}_1,\bar{p}_2,...,\bar{p}_k,...\bar{p}_r). \tag{7.41}$$

For a single particle wave packet we have

$$i\frac{\partial f(s,\bar{x})}{\partial s} = \omega_{\bar{x}} f(s,\bar{x}), \tag{7.42}$$

$$i\frac{\partial g(s,\bar{p})}{\partial s} = \omega_{\bar{p}}\, g(s,\bar{p}). \tag{7.43}$$

Solution of the last equation is

$$g(s,\bar{p}) = \mathrm{e}^{-i\omega_{\bar{p}}s} g(\bar{p}). \tag{7.44}$$

It contains only positive frequencies, because Eq. (7.43) contains the first order derivative with respect to s. We will now show [170] how the wave packet profiles (wave functions) f and g are related to the solution of the

Klein-Gordon equation which contains both positive and negative frequencies. A similar procedure was adopted by Deriglazov [171].

Writing the wave function in Eqs. (7.42) and (7.43) as the sum of the real and imaginary component

$$f = f_R + if_I , \qquad g = g_R + ig_I, \tag{7.45}$$

we obtain

$$\dot{f}_R = \omega_{\bar{x}} f_I , \qquad \dot{f}_I = -\omega_{\bar{x}} f_R, \tag{7.46}$$

$$\dot{g}_R = \omega_{\bar{p}} g_I , \qquad \dot{g}_I = -\omega_{\bar{p}} g_R. \tag{7.47}$$

The system of two equation that have first order derivative of f or g with respect to s, we can write as a system of a single second order equation. This can be achieved by using the left hand side equations (7.46), (7.47), express f_I and g_I according to

$$f_I = \omega_{\bar{x}}^{-1} \dot{f}_R , \qquad \text{and} \qquad g_I = \omega_{\bar{p}}^{-1} \dot{g}_R, \tag{7.48}$$

and insert it into the right hand side equations (7.46), (7.47). So we obtain [171, 173]:

$$\ddot{f}_R + \omega_{\bar{x}}^2 f_R = 0, \tag{7.49}$$

$$\ddot{g}_R + \omega_{\bar{p}}^2 g_R = 0. \tag{7.50}$$

Recalling that $\omega_{\bar{x}}^2 = m^2 + \bar{\partial}_\mu \bar{\partial}^\mu$ and $\omega_{\bar{p}}^2 = m^2 + \bar{p}_\mu \bar{p}^\mu$, we see that Eqs. (7.49), (7.50) are the expressions for the Klein-Gordon equation in position and momentum space, respectively.

In view of (7.45), (7.48), the complex valued wave packet profiles (wave function), f and g, are thus the sums

$$f = f_R + i\omega_{\bar{x}}^{-1} \dot{f}_R, \tag{7.51}$$

$$g = g_R + i\omega_{\bar{p}}^{-1} \dot{g}_R. \tag{7.52}$$

A general solution of Eq. (7.50) is

$$g_R(s, \bar{p}) = A(\bar{p})e^{-i\omega_{\bar{p}}s} + A^*(\bar{p})e^{i\omega_{\bar{p}}s}. \tag{7.53}$$

Using (7.48), we have

$$ig_I = \frac{i}{\omega_{\bar{p}}} \dot{g}_R = Ae^{-i\omega_{\bar{p}}s} - A^* e^{i\omega_{\bar{p}}s}, \tag{7.54}$$

amd thus

$$g(s, \bar{p}) = g_R(s, \bar{p}) + ig_I(s, \bar{p}) = 2A(\bar{p})e^{-i\omega_{\bar{p}}s}. \tag{7.55}$$

Comparison with Eq. (7.46) gives $2A(\bar{p}) = g(\bar{p})$. We have thus explicitly demonstrated that, although occurring in the second order equation (7.50) for the real function g_R, negative frequencies do not occur in the complex valued wave function $g(s, \bar{p})$, and hence also not in the Fourier transformed wave function $f(s, \bar{x})$.

Hence, from a real function $\phi(s, \bar{x}) \equiv f_R(s, \bar{x})$, satisfying the Klein-Gordon equation, we obtain a complex valued function, f, as the sum $f = \phi + i\omega_{\bar{x}}^{-1}\dot{\phi}_R$. Similar holds for the Fourier transformed function $\tilde{\phi} \equiv g_R(s, \bar{p})$, from which we obtain the complex valued function $g = \tilde{\phi} + i\omega_{\bar{p}}^{-1}\dot{\tilde{\phi}}$. But $\dot{\phi}$ is just the field momentum. Therefore, f is proportional to $\frac{1}{\sqrt{2}}\left(\omega_{\bar{x}}^{1/2}\phi + i\omega_{\bar{x}}^{-1/2}\Pi_\phi\right)$, the proportionality factor being $\sqrt{2\omega_{\bar{x}}}$. The single particle state (7.34) can thus be written as

$$|\Psi\rangle = \int d\Sigma \, f(s, \bar{x}) a^\dagger(\bar{x}) |0\rangle$$

$$= \int d\Sigma \sqrt{\frac{2}{\omega_{\bar{x}}}} \frac{1}{\sqrt{2}}\left(\omega_{\bar{x}}^{1/2}\phi + i\omega_{\bar{x}}^{-1/2}\Pi_\phi\right) \frac{1}{\sqrt{2}}\left(\omega_{\bar{x}}^{1/2}\varphi - i\omega_{\bar{x}}^{-1/2}\Pi_\varphi\right)|0\rangle,$$

$$(7.56)$$

which explicitly illustrates that the wave function f and the creation operators $a^\dagger(\bar{x})$ are of the same form, as they should be.

While Φ and the momentum Π_ϕ are classical (c-number) fields, φ and Π_φ are the corresponding quantum fields, i.e., operators. Depending on whether we work in the Schrödinger or Heisenberg picture, either ϕ or φ satisfy the Klein-Gordon equation. For instance, in the Schrödinger picture $\phi = \phi(s, \bar{x})$ satisfies the Klein-Gordon equation (7.49), while $\varphi = \varphi(\bar{x})$ is independent of s (which served the role of time). The opposite holds in the Heisenberg picture.

We have thus elucidated the meaning of the wave function in the Klein-Gordon equation. To a real field ϕ, satisfying the Klein-Gordon equation, there corresponds a complex wave function, f, which is the superposition (7.51) of the real field $\phi \equiv f_R$ and its time derivative (or the s-derivative in in our covariant formulation). The absolute square $|f|^2 = f^*f$ gives the probability density of observing the particle at position $\bar{x}^\mu$.

In the case of a complex Klein-Gordon field, which is a superposition of two independent real fields, $\phi = \phi_1 + i\phi_2$, to each of those fields there corresponds a distinct wave function, namely

$$f_1 = \phi_1 + i\omega_{\bar{x}}^{-1}\dot{\phi}_1 \, , \qquad f_2 = \phi_2 + i\omega_{\bar{x}}^{-1}\dot{\phi}_2. \qquad (7.57)$$

Each of them separately determines the probability density of observing the particle at position $\bar{x}^\mu$, namely, $|f_1|^2$ of particle 1, and $|f_2|^2$ of particle 2.

A generic single particle state of the quantum field theory of a *non Hermitian scalar field* $\varphi = \varphi_1 + i\varphi_2$ is then a generalization of the state (7.56):

$$|\Psi\rangle = \int d\Sigma \left(f_1(s,\bar{x}) a_1^\dagger(\bar{x}) + f_2(s,\bar{x}) a_2^\dagger(\bar{x}) \right) |0\rangle$$

$$= \int d\Sigma \left(\chi_+(s,\bar{x}) b_+^\dagger(\bar{x}) + \chi_-(s,\bar{x}) b_-^\dagger(\bar{x}) \right) |0\rangle, \qquad (7.58)$$

where instead of a single particle creation operator (7.15), we have now two creation operators:

$$a_1^\dagger = \sqrt{\frac{1}{2}} \left(\omega_{\bar{x}}^{1/2} \varphi_1 - i\omega_{\bar{x}}^{-1/2} \dot{\varphi}_1 \right) , \qquad a_2^\dagger = \sqrt{\frac{1}{2}} \left(\omega_{\bar{x}}^{1/2} \varphi_2 - i\omega_{\bar{x}}^{-1/2} \dot{\varphi}_2 \right) . \qquad (7.59)$$

The states with definite charge are created with the operators $b_+^\dagger$ and $b_-^\dagger$, which are linear combinations of the operators $a_1^\dagger$ and $a_2^\dagger$:

$$b_+^\dagger = \frac{a_1^\dagger + ia_2^\dagger}{\sqrt{2}} , \qquad b_-^\dagger = \frac{a_1^\dagger - ia_2^\dagger}{\sqrt{2}}. \qquad (7.60)$$

The corresponding wave packet profiles (wave functions) are[2]:

$$\chi_+ = \frac{f_1 - if_2}{\sqrt{2}} , \qquad \chi_- = \frac{f_1 + if_2}{\sqrt{2}}. \qquad (7.61)$$

Thus

$$|\chi_+|^2 = \frac{1}{2} \left(|f_1|^2 + |f_2|^2 \right) + \frac{i}{2} \left(f_1^* f_2 - f_1 f_2^* \right) \qquad (7.62)$$

determines the probability density of observing at position $\bar{x}^\mu$ a particle with charge $+1$. Similarly,

$$|\chi_-|^2 = \frac{1}{2} \left(|f_1|^2 + |f_2|^2 \right) - \frac{i}{2} \left(f_1^* f_2 - f_1 f_2^* \right) \qquad (7.63)$$

determines the probability density of observing at $\bar{x}^\mu$ a particle with charge -1. The sum is $|\chi_+|^2 + |\chi_-|^2 = |f_1|^2 + |f_2|^2$.

Each of the functions f_1 and f_2 in the composition (7.61) of the states χ_+ and χ_- satisfies the Schrödinger equation (7.42). Therefore we also have

$$i\frac{\partial \chi_+}{\partial s} = \omega_{\bar{x}} \chi_+ , \qquad i\frac{\partial \chi_-}{\partial s} = \omega_{\bar{x}} \chi_-. \qquad (7.64)$$

---

[2]See also Ref. [172].

Throughout the history there has been a lot of confusion about the meaning of a function ϕ that satisfies the Klein-Gordon equation. Interpreting the absolute square $|\phi|^2$ of a complex function as denoting the probability density turned out to be inconsistent. Instead, it was correctly observed that $j_\mu = \frac{i}{2}\left(\phi^*\partial_\mu\phi - \partial_\mu\phi^*\phi\right)$ can be interpreted as a charge current. But as we have shown in this chapter (see also [162, 166, 167, 170]), the probability density can also be defined. However, this is only a part of the story. We also need to define the probability current, what we will do in the following.

Let us first consider the case of *a Hermitian field*, $\varphi = \varphi^\dagger$. The scalar product of a single particle state (7.33) with itself gives

$$\langle\Psi|\Psi\rangle = \int \mathrm{d}\Sigma_{\bar{p}}\, g^*(s,\bar{p})g(s,\bar{p}) = \int \mathrm{d}\Sigma f^*(s,\bar{x})f(s,\bar{x}). \tag{7.65}$$

In terms of the redefined quantities (7.24), (7.36), (7.37), the scalar product reads

$$\langle\Psi|\Psi\rangle = \int \frac{\mathrm{d}\Sigma_{\bar{p}}}{(2\pi)^3 2\omega_{\bar{p}}}\, \tilde{g}^*(s,\bar{p})\tilde{g}(s,\bar{p}) = \int \mathrm{d}\Sigma\, \sqrt{2\omega_{\bar{x}}}\tilde{f}^*(s,\bar{x})\sqrt{2\omega_{\bar{x}}}\tilde{f}(s,\bar{x})$$

$$= \int \mathrm{d}\Sigma\left((\omega_{\bar{x}}\tilde{f}^*)\tilde{f} + \tilde{f}^*\omega_{\bar{x}}\tilde{f}\right) = i\int \mathrm{d}\Sigma\left(\tilde{f}^*\frac{\partial\tilde{f}}{\partial s} - \frac{\partial\tilde{f}^*}{\partial s}\tilde{f}\right). \tag{7.66}$$

In the above equation we performed the partial integration and omitted the surface term, so that

$$\int \mathrm{d}\Sigma f^* f = \int \mathrm{d}\Sigma\sqrt{2\omega_{\bar{x}}}\tilde{f}^*\sqrt{2\omega_{\bar{x}}}\tilde{f} = \int \mathrm{d}\Sigma\left((\omega_{\bar{x}}\tilde{f}^*)\tilde{f} + \tilde{f}^*\omega_{\bar{x}}\tilde{f}\right). \tag{7.67}$$

Then we used the Schrödinger equation for f, (7.42), which, after the substitution $f = \sqrt{2\omega_{\bar{x}}}\tilde{f}$ gives $i\partial\tilde{f}/\partial s = \omega_{\bar{x}}\tilde{f}$.

As $g(s,\bar{p})$ and $f(s,\bar{x})$, also the redefined functions $\tilde{g}(s,\bar{p})$ and $\tilde{f}(s,\bar{x})$ contain only positive frequencies. The expression

$$\rho = f^* f = \sqrt{2\omega_{\bar{x}}}\tilde{f}^*\sqrt{2\omega_{\bar{x}}}\tilde{f} \tag{7.68}$$

is always positive. But the expression

$$\tilde{\rho} = (\omega_{\bar{x}}\tilde{f}^*)\tilde{f} + \tilde{f}^*\omega_{\bar{x}}f = i\left(f^*\frac{\partial\tilde{f}}{\partial s} - \frac{\partial\tilde{f}^*}{\partial s}\tilde{f}\right), \tag{7.69}$$

which differs from $f^* f$ by a total derivative, can be negative, despite that $\tilde{f}$ contains only positive frequencies [162].

If we differentiate the probability density $\rho = f^* f$ with respect to s and use (7.42), we obtain

$$\frac{\partial \rho}{\partial s} = \frac{\partial f^*}{\partial s} f + f^* \frac{\partial f}{\partial s} = i\left((\omega_{\bar{x}} f^*)f - f^* \omega_{\bar{x}} f\right) = \bar{\partial}^\mu \bar{j}_\mu, \tag{7.70}$$

where[3]

$$\bar{j}_\mu = \frac{i}{2m}\left(\bar{\partial}_\mu f^* f - f^* \bar{\partial}_\mu f\right) + \text{higher order terms.} \tag{7.71}$$

The higher order terms in Eq. (7.71) come from the expression $\omega_{\bar{x}} = \sqrt{m^2 + \bar{\partial}_\mu \bar{\partial}^\mu} = m\left(1 + \frac{\bar{\partial}_\mu \bar{\partial}^\mu}{2m^2} - \frac{1}{2.4}\left(\frac{\bar{\partial}_\mu \bar{\partial}^\mu}{m^2}\right)^2 + \frac{1.3}{2.4.6}\left(\frac{\bar{\partial}_\mu \bar{\partial}^\mu}{m^2}\right)^3 - \ldots\right)$.

Because the formalism presented here is covariant, the probability density ρ transforms under Lorentz transformations as a scalar, and the current $\bar{j}_\mu$ as a 4-vector.

Alternatively, if we differentiate $\tilde{\rho}$, defined in Eq. (7.69), we have

$$\frac{\partial \tilde{\rho}}{\partial s} = i\left(\tilde{f}^* \frac{\partial^2 \tilde{f}}{\partial s^2} - \frac{\partial^2 \tilde{f}^*}{\partial s^2} \tilde{f}\right) = i\left(\tilde{f}^* \omega_{\bar{x}}^2 \tilde{f} + (\omega_{\bar{x}}^2 \tilde{f}^*)\tilde{f}\right)$$

$$= i\left(-\tilde{f}^*\left(m^2 + \bar{\partial}_\mu \bar{\partial}^\mu\right)\tilde{f} + \left(m^2 + \bar{\partial}_\mu \bar{\partial}^\mu\right)\tilde{f}^* \tilde{f}\right) = \bar{\partial}_\mu \tilde{j}^\mu, \tag{7.72}$$

where

$$\tilde{j}^\mu = -i\left(\tilde{f}^* \bar{\partial}^\mu \tilde{f} - \bar{\partial}^\mu \tilde{f}^* \tilde{f}\right). \tag{7.73}$$

In the case of *a non Hermitian field* $\varphi \neq \varphi^\dagger$, a generic single particle state is given by Eq. (7.58), which contains two types of creation operators and the corresponding wave functions (that determine wave packet profiles), expressed according to Eqs. (7.60), (7.61).

The scalar product of the state (7.58) with itself gives

$$\langle \Psi | \Psi \rangle = \int d\Sigma \left(f_1^* f_1 + f_2^* f_2\right) = \int d\Sigma \left(\chi_+^* \chi_+ + \chi_-^* \chi_-\right). \tag{7.74}$$

We see that there are two types of probability densities, namely, $\rho_+ = \chi_+^* \chi_+$ for type "+" particles, and $\rho_- = \chi_-^* \chi_-$ for type "−" particles. Analogously to equations (7.70), (7.71), we can define the associated currents $\bar{j}_+^\mu$ and $\bar{j}_-^\mu$.

Redefining the wave function according to (7.37), so that $\chi_\pm = \sqrt{2\omega_{\bar{x}}}\tilde{\chi}_\pm$, we have

$$\langle \Psi | \Psi \rangle = \int d\Sigma \left(\sqrt{2\omega_{\bar{x}}}\,\tilde{\chi}_+^* \sqrt{2\omega_{\bar{x}}}\,\tilde{\chi}_+ + \sqrt{2\omega_{\bar{x}}}\,\tilde{\chi}_-^* \sqrt{2\omega_{\bar{x}}}\,\tilde{\chi}_-\right)$$

---

[3]In Ref. [173] we found that the next higher order terms in the case $n^\mu = (1,0,0,0)$, is $-\frac{imf^*}{2.4}\left(\overleftarrow{\nabla}^3 - \overleftarrow{\nabla}^2\overrightarrow{\nabla} + \overleftarrow{\nabla}\overrightarrow{\nabla}^2 - \overrightarrow{\nabla}^3\right)f$.

$$= \int \mathrm{d}\Sigma \left(\omega_{\bar{x}} \, \tilde{\chi}_+^* \, \tilde{\chi}_+ + \tilde{\chi}_+^* \omega_{\bar{x}} \tilde{\chi}_+ + \omega_{\bar{x}} \, \tilde{\chi}_-^* \, \tilde{\chi}_- + \tilde{\chi}_-^* \omega_{\bar{x}} \tilde{\chi}_- \right). \qquad (7.75)$$

Using the Schrödinger equation (7.64), we find

$$\langle \Psi | \Psi \rangle = \int \mathrm{d}\Sigma \left(\tilde{\chi}_+^* \frac{\partial \tilde{\chi}_+}{\partial s} - \frac{\partial \tilde{\chi}_+^*}{\partial s} \tilde{\chi}_+ + \tilde{\chi}_-^* \frac{\partial \tilde{\chi}_-}{\partial s} - \frac{\partial \tilde{\chi}_-^*}{\partial s} \tilde{\chi}_- \right). \qquad (7.76)$$

Again, in analogy to Eqs. (7.70), (7.71), one can define the associated currents $\tilde{j}_+^{\mu}$ and $\tilde{j}_-^{\mu}$.

7.2.2 The action principle for the wave packet profiles and conserved currents

The equations of motion (7.42) for a single particle wave packet profile, associated with *a Hermitian scalar field*, can be derived from the action[4]

$$I[f, f^*] = \int \mathrm{d}t \, \mathrm{d}^3 \boldsymbol{x} \, \frac{1}{2} \left[i(f^* \dot{f} - \dot{f}^* f) - (\omega_{\boldsymbol{x}} f^*) f - f^8 \omega_b \boldsymbol{x} f \right]. \qquad (7.77)$$

The variation with respect to f and f^* gives

$$
\begin{aligned}
\delta I &= \int \mathrm{d}t \, \mathrm{d}^3 \boldsymbol{x} \left(\frac{\delta I}{\delta f} \delta f + \frac{\delta I}{\delta f^*} \delta f^* \right) \\
&= \int \mathrm{d}t \, \mathrm{d}^3 \boldsymbol{x} \mathrm{d}t' \, \mathrm{d}^3 \boldsymbol{x}' \frac{1}{2} \Big\{ i(f^* \delta f - \delta f^* f) \frac{\mathrm{d}}{\mathrm{d}t} \delta(t - t') \delta^3(\boldsymbol{x} - \boldsymbol{x}') \\
&\quad - (f^* \delta f + \delta f^* f) \delta(t - t') \omega_{\boldsymbol{x}} \delta^3(\boldsymbol{x} - \boldsymbol{x}') \\
&\quad + \left[(i\dot{f} - \omega_{\boldsymbol{x}} f) \delta f^* + (-i\dot{f}^* - \omega_{\boldsymbol{x}} f^*) \delta f \right] \delta(t - t') \delta(\boldsymbol{x} - \boldsymbol{x}') \Big\} (7.78)
\end{aligned}
$$

Taking into account the equation of motion, $i\dot{f} - \omega_{\boldsymbol{x}} f = 0$, performing the partial integration, omitting the surface term, and integrating out the delta functions, we obtain

$$\delta I = \int \mathrm{d}t \, \mathrm{d}^3 \boldsymbol{x} \frac{1}{2} \left[-i \frac{\mathrm{d}}{\mathrm{d}t} (f^* \delta f) + i \frac{\mathrm{d}}{\mathrm{d}t} (\delta f^* f) - \omega_{\boldsymbol{x}} (f^* \delta f) - \omega_{\boldsymbol{x}} (\delta f^* f) \right]. \qquad (7.79)$$

The action (7.77) is invariant under the phase transformations

$$f \longrightarrow \mathrm{e}^{i\alpha} f, \qquad (7.80)$$

---

[4]Now we use the coordinate system in which $n^{\mu} = (1, 0, 0, 0)$, so that $\bar{x}^{\mu} = (0, \boldsymbol{x})$, $s = x^0 \equiv t$, $\mathrm{d}\Sigma = \mathrm{d}^3 \boldsymbol{x}$, $\frac{\partial f}{\partial s} = \frac{\partial f}{\partial t} \equiv \dot{f}$, $\omega_{\bar{x}} = \omega_{\boldsymbol{x}}$.

hence $\delta I = 0$. Inserting the expressions for the infinitesimal variation $\delta f = i\alpha f$ and $\delta f^* = -i\alpha f^*$ into Eq. (7.79), the terms with ω_x cancel out, and we obtain

$$\delta I = \int \mathrm{dt}\, \mathrm{d}^3 x (-i\alpha) \frac{\mathrm{d}}{\mathrm{dt}} (f^* f) = -i\alpha \int \mathrm{d}^3 x\, f^* f \Big|_{t_1}^{t_2} = 0. \qquad (7.81)$$

The conserved generator of the phase transformations (7.80) acting on the action (7.77) is thus

$$G = -i\alpha \int \mathrm{d}^3 x\, f^* f. \qquad (7.82)$$

It is proportional to the conserved probability $P = \int \mathrm{d}^3 x\, f^* f$. Differentiating the probability density $\rho = f^* f$ with respect to time, we obtain the probability current density, as shown in Eqs. (7.70), (7.71).

In the case of a *non Hermitian* $\varphi = \varphi_1 + i\varphi_2$, a generic single particle state (7.34) is given in terms of two wave packets $f_1 = \phi_1 + \omega_x^{-1}\dot{\phi}_1$ and $f_2 = \phi_2 + \omega_x^{-1}\dot{\phi}_2$. They satisfy the action principle

$$I[f_1, f_1^*, f_2, f_2^*] = \int \mathrm{dt}\, \mathrm{d}^3 x \frac{1}{2}\Big[i(f_1^* \dot{f}_1 - \dot{f}_1^*) - (\omega_x f_1^*)f_1 - f_1^* \omega_x f_1$$

$$+ i(f_2^* \dot{f}_2 - \dot{f}_2^*) - (\omega_x f_2^*)f_2 - f_2^* \omega_x f_2 \Big]. \qquad (7.83)$$

There are two distinct types of transformations, involving f_1 and f_2, that leave the above action invariant.

1) One possible transformation is a phase transformation:

$$f_1 \rightarrow e^{i\alpha} f_1 , \qquad f_2 \rightarrow e^{i\alpha} f_2. \qquad (7.84)$$

In view of the relations (7.57), the above transformation is a rotation between ϕ_i and $\omega_x \dot{\phi}_{i1}$ for $i = 1, 2$ (recall that $\phi_1 \equiv f_{1R}$ and $\phi_2 \equiv f_{2R}$:

$$\phi_1 \longrightarrow \cos\alpha\, \phi_1 + \sin\alpha\, \omega_x^{-1}\dot{\phi}_1,$$
$$\omega_x^{-1}\dot{\phi}_1 \longrightarrow -\sin\alpha\phi_1 + \cos\alpha\, \omega_x^{-1}\dot{\phi}_1,$$
$$\phi_2 \longrightarrow \cos\alpha\, \phi_2 + \sin\alpha\, \omega_x^{-1}\dot{\phi}_2,$$
$$\omega_x^{-1}\dot{\phi}_2 \longrightarrow -\sin\alpha\, \phi_2 + \cos\alpha\, \omega_x^{-1}\dot{\phi}_2. \qquad (7.85)$$

A procedure, analogous to that of Eqs. (7.77)–(7.82), now gives the generator

$$G_1 = -i\alpha \int \mathrm{d}^3 x\, (f_1^* f_1 + f_2^* f_2). \qquad (7.86)$$

It contains the probability densities $f_1^* f_1$ and $f_2^* f_2$, associated with the type 1 and type 2 particle, respectively.

2) Another possible transformations is a rotation between ϕ_1 and ϕ_2:

$$\begin{aligned} \phi_1 &\longrightarrow \cos \beta \, \phi_1 + \sin \beta \, \phi_2, \\ \phi_2 &\longrightarrow -\sin \beta \, \phi_1 + \cos \beta \, \phi_2. \end{aligned} \tag{7.87}$$

The infinitesimal transformation being

$$\delta\phi_1 = \beta\phi_2 , \qquad \delta\phi_2 = -\beta\phi_1. \tag{7.88}$$

Written in terms of the complex wave packets $\phi = \phi_1 + i\phi_2$, $\phi^* = \phi_1 - i\phi_2$, the transformation (7.87), (7.88) reads:

$$\phi \longrightarrow e^{i\beta}\phi , \qquad \phi^* \longrightarrow e^{-i\beta}\phi^*, \tag{7.89}$$

$$\delta\phi = i\beta\phi , \quad \delta\phi^* = -i\beta\phi. \tag{7.90}$$

Written in terms of $f_1 = \phi_1 + i\omega_{\boldsymbol{x}}^{-1}\dot\phi_1$ and $f_2 = \phi_2 + i\omega_{\boldsymbol{x}}^{-1}\dot\phi_2$, the transformation (7.87) reads

$$\begin{aligned} f_1 &\longrightarrow \cos \beta \, f_1 + \sin \beta \, f_2, \\ f_2 &\longrightarrow -\sin \beta \, f_1 + \cos \beta \, f_2, \end{aligned} \tag{7.91}$$

$$\begin{aligned} f_1^* &\longrightarrow \cos \beta \, f_1^* + \sin \beta \, f_2^*, \\ f_2^* &\longrightarrow -\sin \beta \, f_1^* + \cos \beta \, f_2^*, \end{aligned} \tag{7.92}$$

$$\delta f_1 = \beta f_2 , \qquad \delta f_2 = -\beta f_1, \tag{7.93}$$

$$\delta f_1^* = \beta f_2^* , \qquad \delta f_2^* = -\beta f_1^*. \tag{7.94}$$

The action (7.83) is equivalent to the action

$$I[\phi_1, \phi_2] = \frac{1}{2} \int dt \, d^3\boldsymbol{x} \left(\dot\phi_1^2 + \dot\phi_2^2 - \phi_1\omega_{\boldsymbol{x}}^2\phi_1 - \phi_1\omega_{\boldsymbol{x}}^2\phi_2 \right). \tag{7.95}$$

It variation is

$$\delta I = \int dt \, d^3\boldsymbol{x} \left(\frac{\delta I}{\delta\phi_1}\delta\phi_1 + \frac{\delta I}{\delta\phi_2}\delta\phi_2 \right). \tag{7.96}$$

Let us take the transformation (7.88). Then the procedure, analogous to Eqs. (7.77)–(7.82) gives

$$\delta I = -\beta \int dt \, d^3\boldsymbol{x} \, \frac{d}{dt} \left(\dot\phi_1\phi_2 - \dot\phi_2\phi_1 \right)$$

$$= -\beta \int d^3x \left(\dot{\phi}_1 \phi_2 - \dot{\phi}_2 \phi_1 \right) \Big|_{t_1}^{t_2}. \tag{7.97}$$

Again the terms with $\omega_{\boldsymbol{x}}$ have canceled out.

The conserved generator of the transformations (7.87) acting on the action (7.95) is thus

$$G_2 = -\beta \int d^3x \left(\dot{\phi}_1 \phi_2 - \dot{\phi}_2 \phi_1 \right). \tag{7.98}$$

It contains the charge density $\dot{\phi}_1 \phi_2 - \dot{\phi}_2 \phi_1$. The charge current density is obtained from

$$\frac{d}{dt} \left(\dot{\phi}_1 \phi_2 - \dot{\phi}_2 \phi_1 \right) = \ddot{\phi}_1 \phi_2 - \ddot{\phi}_2 \phi_1 = -\partial_i \left(\partial^i \phi_1 \, \phi_2 - \partial^i \phi_2 \, \phi_1 \right), \tag{7.99}$$

where in the last step the equations of motion

$$\ddot{\phi}_1 + \omega_{\boldsymbol{x}} \phi_1 = 0 \;, \quad \text{and} \quad \ddot{\phi}_2 + \omega_{\boldsymbol{x}} \phi_2 = 0 \tag{7.100}$$

have been used.

The conserved charge and current can be obtained also by rewriting the action (7.95) into the following equivalent form

$$I[\phi_1, \phi_2] = \frac{1}{2} \int dt \, d^3x \left(\dot{\phi}_1^2 + \dot{\phi}_2^2 + \partial_i \phi_1 \partial^i \phi_1 + \partial_i \phi_2 \partial^i \phi_2 - m^2 \phi_1^2 - m^2 \phi_2^2 \right), \tag{7.101}$$

and employing the familiar Noether's relation

$$\delta I[\phi_1, \phi_2] = \int dt \, d^3x \left[\frac{d}{dt} \left(\frac{\partial \mathcal{L}}{\partial \dot{\phi}_1} \delta \phi_1 + \frac{\partial \mathcal{L}}{\partial \dot{\phi}_2} \delta \phi_2 \right) \right.$$

$$\left. + \partial_i \left(\frac{\partial \mathcal{L}}{\partial \partial_i \phi_1} \delta \phi_1 + \frac{\partial \mathcal{L}}{\partial \partial_i \phi_2} \delta \phi_2 \right) \right], \tag{7.102}$$

which now gives

$$\delta I = \beta \int dt \, d^3x \left[\frac{d}{dt} \left(\dot{\phi}_1 \phi_2 - \dot{\phi}_2 \phi_1 \right) + \partial_i \left(\partial^i \phi_1 \, \phi_2 - \partial^i \phi_2 \, \phi_1 \right) \right]. \tag{7.103}$$

Written in terms of ϕ, ϕ^*, the action (7.101) reads

$$I[\phi, \phi^*] = \frac{1}{2} \int dt \, d^3x \left(\dot{\phi}^* \dot{\phi} + \partial_i \phi^* \partial^i \phi - m^2 \phi^* \phi \right). \tag{7.104}$$

Taking the transformations (7.90) and using an expression, analogous to (7.102), in which ϕ_1, ϕ_2 are replaced by ϕ, ϕ^*, we obtain

$$\delta I[\phi, \phi^*] = -i\beta \int dt \, d^3x \left[\frac{d}{dt} \left(\dot{\phi}^* \phi - \phi^* \dot{\phi} \right) + \partial_i \left(\partial^i \phi^* \, \phi - \phi^* \partial^i \phi \right) \right]. \tag{7.105}$$

This is the usual expression that gives the conserved charge and current.

We can as well start from the action (7.83), which involves the complex wave packets f_1 and f_2. Using the infinitesimal transformations (7.93), (7.94), we obtain

$$\delta I[f_1, f_2] = i\beta \int dt\, d^3x \, \frac{d}{dt} \left(f_1^* f_2 - f_2^* f_1 \right), \qquad (7.106)$$

where again the terms with ω_x have cancelled out.

The charge density is now expressed as

$$\rho_e = -i(f_1^* f_2 - f_2^* f_1) = \chi_+^* \chi_+ - \chi_-^* \chi_-, \qquad (7.107)$$

where the relation (7.61) between f_1, f_2 and χ_+, χ_- has been used.

We have seen that there are two distinct types of the transformations between the fields occurring in the action (7.83), or its equivalent forms (7.95), (7.101) and (7.104). Both types leave the action invariant, and hence lead to the corresponding conserved Noether's charges. In the case of the transformation which involves a field and its time derivative, the conserved "charge" is the probability of observing the particle. In the case of the transformation which rotates one field into the other, the conserved charge is the electric (or equivalent) charge.

7.2.3   *Working with field operators*

A generic single particle state is given by equations (7.58), which we now write in a more compact form:

$$|\Psi\rangle = \int d^3x \, f^i(x) a_i^\dagger(x)|0\rangle \,, \qquad i = 1, 2, \qquad (7.108)$$

where either $f^i(x)$ or $a_i^\dagger(x)$ can depend on time. If the state is determined at time $t = 0$, then the state at a time t is

$$|\Psi(t)\rangle = e^{-iHt}|\psi(0)\rangle = \int d^3x \, f^i(0, x) e^{-iHt} a_i^\dagger(0, x) e^{iHt} e^{-iHt}|0\rangle. \quad (7.109)$$

Assuming that the vacuum is invariant under time translations, $e^{-iHt}|0\rangle = |0\rangle$, we have

$$|\Psi(t)\rangle = \int d^3x \, f^i(0, x) a_i^\dagger(t, x)|0\rangle, \qquad (7.110)$$

where

$$a_i^\dagger(t, x) = e^{-iHt} a_i^\dagger(0, x) e^{iHt}. \qquad (7.111)$$

Using the expansion $e^{-iHt} = 1 - iHt + ...$, which gives $a_i^\dagger(t, \boldsymbol{x}) = a_i^\dagger(0, \boldsymbol{x}) + i[a_i^\dagger(0, \boldsymbol{x}), H] + ...$, we obtain

$$\frac{\partial}{\partial t} a_i^\dagger = i[a_i^\dagger, H]. \tag{7.112}$$

Omitting the zero point term we find that the Hamilton operator (7.19) in the Lorentz frame in which $n^\mu = (1, 0, 0, 0)$, and extended to two fields, is

$$H = \int \mathrm{d}^3 \boldsymbol{x} \, a_i^\dagger(\boldsymbol{x}) \omega_{\boldsymbol{x}} a^i(\boldsymbol{x}), \tag{7.113}$$

and we have

$$[a_i^\dagger(\boldsymbol{x}), H] = -\omega_{\boldsymbol{x}} a_i^\dagger(\boldsymbol{x}). \tag{7.114}$$

The Schrödinger equation then gives

$$i\frac{\partial|\Psi\rangle}{\partial t} = H|\Psi\rangle = \int \mathrm{d}^3 \boldsymbol{x} \, f^i(\boldsymbol{x})(-1)[a_i^\dagger, H]|0\rangle$$

$$= \int \mathrm{d}^3 \boldsymbol{x} \, f^i(\boldsymbol{x}) \omega_{\boldsymbol{x}} a_i^\dagger(\boldsymbol{x})|0\rangle = \int \mathrm{d}^3 \boldsymbol{x} \, (\omega_{\boldsymbol{x}} f^i(\boldsymbol{x})) a_i^\dagger(\boldsymbol{x})|0\rangle \tag{7.115}$$

In evaluating the left hand sided of the latter equation we can assume that

1) f^i is time dependent and $a_i^\dagger$ constant, in which case we have

$$i\frac{\partial|\Psi\rangle}{\partial t} = \int \mathrm{d}^3 \boldsymbol{x} \frac{\partial f^i}{\partial t} a_i^\dagger|0\rangle. \tag{7.116}$$

Using (7.115) we thus obtain

$$i\frac{\partial f^i}{\partial t} = \omega_{\boldsymbol{x}} f^i. \tag{7.117}$$

2) f^i is constant and $a_i^\dagger$ time dependent, so that

$$i\frac{\partial|\Psi\rangle}{\partial t} = \int \mathrm{d}^3 \boldsymbol{x} f^i \frac{\partial a_i^\dagger}{\partial t}|0\rangle. \tag{7.118}$$

Using (7.115) we now obtain

$$i\frac{\partial a_i^\dagger}{\partial t} = -[a_i^\dagger, H] = \omega_{\boldsymbol{x}} a_i^\dagger(\boldsymbol{x}). \tag{7.119}$$

In case 1 the components $f^i(\boldsymbol{x})$ of a state vector change with time, while the basis vectors $|\boldsymbol{x}\rangle = a_i^\dagger(\boldsymbol{x})|0\rangle$ are constant. In case 2 it is the opposite: the components are constant and the basis vectors change with time.

We have analyzed the action of the time translation (evolution) operator e^{-iHt} on a generic single particle state (7.108). Analogous can be done for any other operator that transforms a state according to

$$|\Psi\rangle \longrightarrow |\Psi'\rangle = e^{-i\alpha Q}|\Psi\rangle = \int d^3x \, f^i e^{-i\alpha Q} a_i^\dagger e^{i\alpha Q}|0\rangle, \qquad (7.120)$$

which means that

$$a_i^\dagger = a^{-i\alpha Q} a_i^\dagger e^{i\alpha Q}, \qquad (7.121)$$

where Q is the generator of the transformation.

Let us assume that Q is given by the following generic quadratic form

$$Q = \int d^3x' \, d^3x'' \, Q^{ij}(x', x'') a_i^\dagger(x') a_j(x''). \qquad (7.122)$$

Some particular cases are:

(i)

$$Q^{ij}(x', x'') = \delta^{ij}\omega_x'\delta^3(x' - x''), \qquad (7.123)$$

which gives

$$Q = \int d^3x \, \delta^{ij} a_i^\dagger(x)\omega_x a_j(x) = H, \qquad (7.124)$$

that is, the Hamilton operator.

(ii)

$$Q^{ij}(x', x'') = \delta^{ij}\delta^3(x' - x''), \qquad (7.125)$$

$$Q = \int d^3x \, \delta^{ij} a_i^\dagger(x) a_j(x), \qquad (7.126)$$

which is the number operator for type 1 and type 2 particles.

(iii)

$$Q^{ij}(x', x'') = i\epsilon^{ij}\delta^3(x' - x''), \qquad (7.127)$$

$$Q = i \int d^3x \, \epsilon^{ij} a_i^\dagger(x) a_j(x), \qquad (7.128)$$

which, as we will see is the charge operator.

For an infinitesimal α, the transformation (7.120) becomes

$$\delta|\Psi\rangle = \alpha \int d^3x \, f^i(x) i[a_i^\dagger, Q]|0\rangle = \alpha \int d^3x \, f^i \delta a_i, \qquad (7.129)$$

where

$$\delta a_i^\dagger(x) = i\alpha[a_i^\dagger(x), Q] = -i\alpha \int d^3x' Q^j{}_i(x', x) a_j^\dagger(x'). \qquad (7.130)$$

For the particular cases (i), (ii), (iii) we have, respectively,

$$(i) \qquad \delta a_i^\dagger = -i\alpha \omega_x a_i^\dagger, \qquad \delta a_i = i\alpha \omega_x a_i, \qquad (7.131)$$

$$(ii) \qquad \delta a_i^\dagger = -i\alpha a_i^\dagger, \qquad \delta a_i = i\alpha a_i, \qquad (7.132)$$

$$(iii) \qquad \delta a_i^\dagger = \alpha \epsilon_i{}^j a_j^\dagger, \qquad \delta a_i = \alpha \epsilon_i{}^j a_j. \qquad (7.133)$$

The expectation values of Q in a single particle state (7.108) is

$$\langle Q \rangle = \int d^3x' \, d^3x'' \, f^{*i}(x') Q_{ij}(x', x'') f^j(x''), \qquad (7.134)$$

which for the particular cases (i), (ii) and (iii) of Eqs. (7.123), (7.125) and (7.127), gives

(i) the expected energy,

$$\langle Q \rangle = \int d^3x \, f^{*i} \omega_x f^j(x) \delta_{ij} = \langle H \rangle, \qquad (7.135)$$

(ii) the probability,

$$\langle Q \rangle = \int d^3x \, f^{*i} f^j(x) \delta_{ij} = \int d^3x \, (\chi_+^* \chi_+ + \chi_-^* \chi_-), \qquad (7.136)$$

(iii) the expected charge,

$$\langle Q \rangle = \int d^3x \, f^{*i} f^j(x) (-i) \epsilon_{ij} = \int d^3x \, (\chi_+^* \chi_+ - \chi_-^* \chi_-). \qquad (7.137)$$

The corresponding finite transformation for the case (i) is

$$a'_i^\dagger = e^{-i\alpha\omega_x} a_i^\dagger, \qquad (7.138)$$

where α is a time interval.

For the case (ii), the finite transformation is

$$a'_i^\dagger = e^{-i\alpha} a_i^\dagger. \qquad (7.139)$$

Recalling Eq. (7.59), we see that this is just a rotation between φ_i and $\omega_x^{-1}\dot\varphi_i$. $i = 1, 2$. Roughly speaking, this is a rotation that mixes a field, say φ_1 and its time derivative $\dot\varphi_1$ (multiplied by ω_x^{-1}). In the alternative basis (7.60), written now as $(b_+^\dagger, b_-^\dagger) \equiv (b_1^\dagger, b_2^\dagger)$, the transformation reads

$$b'_i^\dagger = e^{-i\alpha} b_i^\dagger, \qquad i = 1, 2. \qquad (7.140)$$

For the case (iii), the finite transformation in the basis $a_i^\dagger$ is

$$a'_1^\dagger = \cos\alpha \, a_1^\dagger + \sin\alpha \, a_2^\dagger,$$
$$a'_2^\dagger = -\sin\alpha \, a_1^\dagger + \cos\alpha \, a_2^\dagger, \qquad (7.141)$$

which is just a rotation that mixes type 1 and type 2 fields. Expressed in terms of the basis $b_i^\dagger$, it reads

$$b'_1^\dagger = e^{-i\alpha} b_1^\dagger, \qquad (7.142)$$

$$b'_2^\dagger = e^{i\alpha} b_2^\dagger. \qquad (7.143)$$

7.2.4   The action principle for creation operators and conserved currents

The equation of motion (7.119) for the creation operators $a_i^\dagger$ follows from the action

$$I = \frac{1}{2} \int dt \left[-i \left(a_{i(x)}^\dagger \dot{a}^{i(x)} - \dot{a}_{i(x)} a^{i(x)} \right) \right.$$
$$\left. - \left(\omega_{i(x)j(x')} a^{\dagger i(x)} \right) a^{j(x)} - a^{i(x)} \omega_{i(x)j(x')} a^{j(x')} \right]. \quad (7.144)$$

Here we extend the Einstein summation convention to integration and write

$$a_{i(x)} \equiv a_i(x), \qquad \omega_{i(x)j(x')} \equiv \delta_{ij} \omega_x \delta(x - x'). \quad (7.145)$$

Performing an infinitesimal variation of the action (7.144) and taking into account the equations of motion

$$-i\dot{a}_{i(x)} = \omega_{i(x)j(x')} a^{i(x')}, \qquad -i\dot{a}_{i(x)}^\dagger = \omega_{i(x)j(x')} a^{\dagger i(x')}, \quad (7.146)$$

we obtain

$$\delta I = \frac{1}{2} \int dt \left[-i \left(a_{i(x)}^\dagger \delta \dot{a}^{i(x)} - \delta \dot{a}_{i(x)} a^{i(x)} \right) \right.$$
$$\left. - \left(\omega_{i(x)j(x')} \delta a^{\dagger i(x)} \right) a^{j(x)} - a^{i(x)} \omega_{i(x)j(x')} \delta a^{j(x')} \right]. \quad (7.147)$$

As usual, we interchange variation and derivative, and write $\delta \dot{a}^{i(x)} = \frac{d}{dt} \delta a^{i(x)}$. The infinitesimal transformation (7.130) which, together with its hermitian conjugate, now reads

$$\delta a_{i(x)}^\dagger = -i\alpha [Q, a_{i(x)}^\dagger] = -i\alpha Q^{j(x')}{}_{i(x)} a_{j(x)}^\dagger, \quad (7.148)$$

$$\delta a_{i(x)} = i\alpha Q^{\dagger j(x')}{}_{i(x)} a_{j(x)}, \quad (7.149)$$

where $Q^{i(x)j(x')} \equiv Q^{ij}(x, x')$, its indices being raised and lowered by the diagonal metric $\delta_{i(x)j(x')} \equiv \delta_{ij} \delta^3(x - x')$. Using (7.148), (7.149), we obtain

$$\delta I = \frac{1}{2} \int dt \left[(-i) \left(a^\dagger{}_{i(x)} \frac{d}{dt} \left(i\alpha Q^{\dagger j(x')i(x)} a_{j(x')} \right) \right. \right.$$
$$-\frac{d}{dt} \left(-i\alpha Q^{j(x')i(x)} a^\dagger{}_{j(x')} \right) a_{i(x)} \right) - \omega_{i(x)j(x')} (-i\alpha) Q^{k(x'')j(x')} a^\dagger{}_{k(x'')} a^{i(x)}$$
$$\left. - a^{\dagger i(x)} \omega_{i(x)j(x')} (i\alpha) Q^{\dagger k(x'')j(x')} a^\dagger{}_{k(x'')} \right]. \quad (7.150)$$

Regardless of whether we consider the case (7.125) in which

$$Q^{i(x)j(x')} = Q^{j(x')i(x)} = \delta^{ij} \delta^3(x - x') = Q^{\dagger i(x)j(x')}, \quad (7.151)$$

or the case (7.127) in which

$$Q^{i(\boldsymbol{x})j(\boldsymbol{x}')} = -Q^{j(\boldsymbol{x}')i(\boldsymbol{x})} = i\epsilon^{ij}\delta^3(\boldsymbol{x} - \boldsymbol{x}') = -Q^{\dagger i(\boldsymbol{x})j(\boldsymbol{x}')}, \qquad (7.152)$$

we obtain

$$\delta I = \alpha \int \mathrm{d}t \, \frac{\mathrm{d}}{\mathrm{d}t} \left(a^{\dagger}{}_{ij(\boldsymbol{x})} Q^{i(\boldsymbol{x})j(\boldsymbol{x}'} a_{j(\boldsymbol{x})} \right). \qquad (7.153)$$

If the action is invariant under the transformation (7.148), (7.149), it must be

$$\frac{\mathrm{d}Q}{\mathrm{d}t} = 0, \qquad (7.154)$$

where

$$Q = a^{\dagger}{}_{ij(\boldsymbol{x})} Q^{i(\boldsymbol{x})j(\boldsymbol{x}'} a_{j(\boldsymbol{x})}, \qquad (7.155)$$

which means that the generator Q is conserved. Because it satisfies the Heisenberg equations of motion

$$\frac{\mathrm{d}Q}{\mathrm{d}t} = -i[Q, H], \qquad (7.156)$$

the operator Q is conserved if it commutes with the Hamiltonian. This is indeed the case for the generator (7.155) and the Hamilton operator

$$H = a^{\dagger i(\boldsymbol{x})} \omega_{i(\boldsymbol{x})j(\boldsymbol{x}')} a^{j(\boldsymbol{x}')}, \qquad (7.157)$$

which is the tensor notation prescription of (7.113) for $\omega_{i(\boldsymbol{x})j(\boldsymbol{x}')} = \omega_{\boldsymbol{x}}\delta_{ij}\delta^3(\boldsymbol{x} - \boldsymbol{x}')$.

Which generator, and hence which transformation and what kind of charge, we happen to consider depends on choice of the matrix $Q^{i(\boldsymbol{x})j(\boldsymbol{x}'}$. Besides the choice (7.123) which gives the generator of time translations (the Hamilton operator), and another choice which gives the generator of spatial translations (the momentum operator), of special interest here are choices (7.151) and (7.152). The former one generates phase transformations (7.84), its expectation value being the probability (7.136), whilst the latter one generates rotations between $a_1^{\dagger}(\boldsymbol{x})$ and $a_2(\boldsymbol{x})$, which can be written as a phase transformation on the basis $b_i^{\dagger} \equiv (b_+^{\dagger}, b_-^{\dagger})$, related to the basis $a_i^{\dagger}$ according to Eq. (7.60). Whilst the choice (7.151) corresponds to the probability, the choice (7.152) corresponds to a charge (e.g., the electric charge). The existence of those two different conserved generators, namely the probability and the charge, has escaped from being clearly and unambiguously observed in the literature on quantum field theory.

7.2.5 *The action principle for states*

The Schrödinger equation can be derived from the action

$$I = \frac{1}{2} \int dt \left[i \langle \Psi | \frac{\partial}{\partial t} | \Psi \rangle - \left(\frac{\partial}{\partial t} \langle \psi | \right) | \Psi \rangle - (H \langle \Psi |) | \psi \rangle - \langle \Psi | (H | \Psi \rangle) \right].$$
(7.158)

Variation with respect to $\langle \Psi |$ gives[5]

$$i \frac{\partial}{\partial t} | \Psi \rangle = H | \Psi \rangle.$$
(7.159)

For a single particle states (7.108) and the Hamiltonian (7.113) we have

$$i \frac{\partial}{\partial t} \int d^3 x \, f^i(\boldsymbol{x}) a_i^\dagger(\boldsymbol{x}) |0\rangle = \int d^3 x f^i(\boldsymbol{x}) \omega_{\boldsymbol{x}} a_i^i(\boldsymbol{x}) |0\rangle$$

$$= \int d^3 x \, (\omega_{\boldsymbol{x}} f^i(\boldsymbol{x})) a_i^i(\boldsymbol{x}) |0\rangle.$$
(7.160)

In the last step we, as usually, omitted the surface term. Therefore, the operator $\omega_{\boldsymbol{x}} = \sqrt{m^2 - \nabla_{\boldsymbol{x}}^2}$ can act either on $a_i^\dagger(\boldsymbol{x})$ or $f^i(\boldsymbol{x})$.

Depending on which one, f^i or $a_i^\dagger$, we assign time dependence, we obtain either

$$i \frac{\partial f^i}{\partial t} = \omega_{\boldsymbol{x}} f^i,$$
(7.161)

or

$$i \frac{\partial a_i^\dagger}{\partial t} = \omega_{\boldsymbol{x}} a_i^\dagger,$$
(7.162)

i.e., the two cases already discussed in Sec. 3.3.

From the differential equation (7.159) which has the first order time derivative we can pass to the second order equation by splitting

$$|\Psi\rangle = |\Psi\rangle_R + i |\Psi\rangle_I,$$
(7.163)

where

$$|\Psi\rangle_R = \int d^3 x \, f_R^i a_i^\dagger |0\rangle, \quad |\Psi\rangle_I = \int d^3 x f_I^i a_i^\dagger |0\rangle.$$
(7.164)

Inserting (7.164) into the Schrödinger equation (7.159), we obtain two equations, one for the real and one for the imaginary part of $|\Psi\rangle$:

$$\frac{\partial |\Psi\rangle_R}{\partial t} = H |\Psi\rangle_I,$$
(7.165)

---

[5]Variation with respect to $|\Psi\rangle$ gives

$$i \frac{\partial}{\partial t} \langle \Psi | = H \langle \Psi | \equiv \langle \Psi | \overleftarrow{H}.$$

$$\frac{\partial |\Psi\rangle_I}{\partial t} = -H|\Psi\rangle_R. \qquad (7.166)$$

The latter two first order equations can be written as a single second order equation

$$\frac{\partial^2 |\Psi\rangle_R}{\partial t^2} + H^2 |\Psi\rangle_R = 0. \qquad (7.167)$$

Using the expression (7.157) for H, we obtain for the second term

$$H^2 |\Psi\rangle_R = \int d^3 x \omega_x^2 f_R^i) a_i^\dagger |0\rangle = \int d^3 dx f_R^i \omega_x^2 a_i^\dagger |0\rangle. \qquad (7.168)$$

Depending on whether we assign time dependence to f_R^i or $a_i^\dagger$, the second order equation (7.167) can be written in two different ways:

$$I. \quad \int d^3 x \left(\ddot{f}_R^i + \omega_x^2 f_R^i \right) a_i^\dagger(x)|0\rangle = 0, \qquad (7.169)$$

which gives

$$\ddot{f}_R^i + \omega_x^2 f_R^i = 0, \qquad (7.170)$$

and

$$II. \quad \int d^3 x \, f_R^i \left(\ddot{a}_i^\dagger + \omega_x^2 a_i^\dagger \right) |0\rangle = 0, \qquad (7.171)$$

which gives

$$\ddot{a}_i^\dagger + \omega_x^2 a_i^\dagger = 0. \qquad (7.172)$$

The case I corresponds to the equation of motion (7.49) for the wave function, considered in Sec. 7.2.1, where it was shown that instead of the first order equation for a complex function $f = f_R + i f_I$, we can have a second a second order equation for a real function $f_R \equiv \phi$.

The case II corresponds to the equation of motion (7.119) for the creation operator. Namely, differentiating the left and and the right side of Eq. (7.119) with respect to time, we obtain $i\ddot{a}_i^\dagger = \omega_x \dot{a}_i^\dagger$. Next, inserting $\dot{a}^\dagger = -i\omega_x a_i^\dagger$, gives the second order equation (7.172). Now recall Eq. (7.14), which says that

$$a_i^\dagger = \sqrt{\frac{\omega_x}{2}} \left(\varphi_i - i\omega_x^{-1} \dot{\varphi}_i \right), \qquad (7.173)$$

i.e.,

$$a_i^\dagger = a_{iR} - i a_{iI}, \qquad (7.174)$$

where

$$a_{iR} = \sqrt{\frac{\omega_{\boldsymbol{x}}}{2}}\,\varphi_i\,, \quad \text{and} \quad a_{iI} = \frac{1}{\sqrt{2\omega_{\boldsymbol{x}}}}\,\dot{\varphi}_i. \tag{7.175}$$

Inserting the expression (7.173) into equation (7.172), we obtain the following equations of motion for the real and imaginary part of $a_I^\dagger$, respectively:

$$\ddot{\varphi}_i + \omega_{\boldsymbol{x}}^2 \varphi_i = 0, \tag{7.176}$$

$$\dddot{\varphi}_i + \omega_{\boldsymbol{x}}^2 \dot{\varphi}_i = 0. \tag{7.177}$$

The first one is just the equation of motion for the quantum field $\varphi_i(t, \boldsymbol{x})$. The second equation is a trivial consequence of the first one, and thus redundant.

Instead of differentiating Eq. (7.119), we can write it as a system of two sets of equations, one for $a_{iR}^\dagger$ and the other one for $a_{iI}^\dagger$:

$$\dot{a}_{iR}^\dagger = \omega_{\boldsymbol{x}} a_{iI}\,, \qquad \dot{a}_{iI}^\dagger = -\omega_{\boldsymbol{x}} a_{iR}, \tag{7.178}$$

which gives

$$\ddot{a}_{iR}^\dagger + \omega_{\boldsymbol{x}}^2 a_{iR}^\dagger = 0\,, \quad \text{and} \quad \ddot{a}_{iI}^\dagger + \omega_{\boldsymbol{x}}^2 a_{iI}^\dagger = 0. \tag{7.179}$$

Using (7.175), we obtain equations (7.176) and (7.177).

7.2.6 *The scalar product between states at different times: The propagator*

Usage of wave packets is not customary in quantum field theory, and especially not of wave packets in position space. In this respect Thomas Blomberg [174] wrote:

> Space localization is stepmotherly left undiscussed in texts on quantum field theory where the treatment is limitted to scattering of plane waves. This might seem sufficient for calculating S-matrix in high energy particle physics. ... A proper discussion of finite space localization needs a discussion of wave-packets and their approximate localization (confidence estimates) to finite space regions. The instantaneous space localizatin and on them ultimately based description of the actual phenomena is only defined relative a special inertial system (the laboratory system). This does not however exclude the use of relativistic dynamics.

In the following quote from Ref. [166] the relation between the usual propagator considered in quantum field theory and the formalism of wave packet states is clarified:

The scalar product of two states (7.34) at different times can be expressed as

$$\langle \Psi_2(t')|\Psi_1(t)\rangle = \langle 0| \int a(\boldsymbol{x}')f_2^*(t',\boldsymbol{x}')\mathrm{d}^3\boldsymbol{x}'\mathrm{d}^3\boldsymbol{x}\, f_1(t,\boldsymbol{x})a^\dagger(\boldsymbol{x})|0\rangle$$

$$= \langle 0| \int \tilde{a}(\boldsymbol{x}')2\omega_{\boldsymbol{x}'}\tilde{f}_2^*(t',\boldsymbol{x}')\mathrm{d}^3\boldsymbol{x}'\mathrm{d}^3\boldsymbol{x}\, 2\omega_{\boldsymbol{x}}\tilde{f}_1(t,\boldsymbol{x})\tilde{a}^\dagger(\boldsymbol{x})|0\rangle,$$

(7.180)

where $\omega_{\boldsymbol{x}} \equiv \sqrt{m^2 - \nabla^2}$ and

$$\tilde{f}(t,\boldsymbol{x}) = (2\omega_{\boldsymbol{x}})^{-1/2} f(t,\boldsymbol{x}).$$

(7.181)

Using the Schrödinger equation (7.42), we have

$$\tilde{f}(t,\boldsymbol{x}) = \mathrm{e}^{-i\omega_{\boldsymbol{x}}t}\tilde{f}(0,\boldsymbol{x}),$$

(7.182)

$$f(t,\boldsymbol{x}) = \mathrm{e}^{-i\omega_{\boldsymbol{x}}t}f(0,\boldsymbol{x}).$$

(7.183)

The initial and final wave packet profiles $f_{1,2}(0,\boldsymbol{x})$ or $\tilde{f}_{1,2}(0,\boldsymbol{x})$ are arbitrary. Let us consider two choices:

$(i) \qquad f_1(0,\boldsymbol{x}) = \delta(\boldsymbol{x}-\boldsymbol{x}_0)\,, \quad f_2(0,\boldsymbol{x}') = \delta(\boldsymbol{x}'-\boldsymbol{x}_0').$ (7.184)

Then we obtain

$$\langle \Psi_2(t')|\Psi_1(t)\rangle = \langle 0|\mathrm{e}^{i\omega_{\boldsymbol{x}_0'}t'}a(\boldsymbol{x}_0')\mathrm{e}^{-i\omega_{\boldsymbol{x}_0}t}a^\dagger(\boldsymbol{x}_0)|0\rangle$$

$$= \mathrm{e}^{i\omega_{\boldsymbol{x}_0}(t'-t)}\delta(\boldsymbol{x}_0' - \boldsymbol{x}_0),$$

(7.185)

which is the Green function $G(t',\boldsymbol{x}';t,\boldsymbol{x})$.

$(ii) \qquad 2\omega_{\boldsymbol{x}}\tilde{f}_1(0,\boldsymbol{x}) = \delta(\boldsymbol{x} - \boldsymbol{x}_0)\,, \quad 2\omega_{\boldsymbol{x}}'\tilde{f}_2(0,\boldsymbol{x}') = \delta(\boldsymbol{x}' - \boldsymbol{x}_0'),$

(7.186)

then

$$\langle \Psi_2(t')|\Psi_1(t)\rangle = \langle 0|\mathrm{e}^{i\omega_{\boldsymbol{x}_0'}t'}\tilde{a}(\boldsymbol{x}_0')\mathrm{e}^{-i\omega_{\boldsymbol{x}_0}t}\tilde{a}^\dagger(\boldsymbol{x}_0)|0\rangle$$

$$= \mathrm{e}^{i\omega_{\boldsymbol{x}_0}(t'-t)}\frac{1}{2\omega_{\boldsymbol{x}_0}}\delta(\boldsymbol{x}_0' - \boldsymbol{x}_0),$$

(7.187)

where we have used $[\tilde{a}(\boldsymbol{x}'),\tilde{a}^\dagger(\boldsymbol{x})] = (1/(2\omega_{\boldsymbol{x}}))\delta(\boldsymbol{x}' - \boldsymbol{x})$. Because $\tilde{a}^\dagger(\boldsymbol{x}) \equiv \varphi^+(0,\boldsymbol{x})$, $\varphi^+(t,\boldsymbol{x}) = \mathrm{e}^{-i\omega_{\boldsymbol{x}}t}\varphi(0,\boldsymbol{x})$, we can write Eq. (7.187) in the form $(x \equiv (t,\boldsymbol{x}))$

$$\langle \Psi_2(t')|\Psi_1(t)\rangle = \langle 0|\varphi(x')\varphi^+(x)|0\rangle\,, \quad t' > t.$$

(7.188)

If we do not impose the condition $t' > t$, then the right hand side of Eq. (7.188) can be written in terms of the time ordered product $\langle 0|T\varphi(x')\varphi(x)|0\rangle$, which is the usual QFT propagator.

Both propagators, (7.185) and (7.187) (i.e., (7.188)), are special cases of the scalar product (7.180)

In the case (i), the initial and final wave packet profiles are localized according to (7.184). This is the localization studied in this paper. The initial, and analogously the final, state are then of the form

$$|\Psi(0)\rangle = \int \mathrm{d}^3x\, f(0,\boldsymbol{x}) a^\dagger(\boldsymbol{x})|0\rangle = a^\dagger(\boldsymbol{x}_0)|0\rangle \equiv |\boldsymbol{x}_0\rangle, \quad (7.189)$$

and the scalar product (7.180) gives (7.185), which can be written as

$$G(t',\boldsymbol{x}';t,\boldsymbol{x}) = \langle \boldsymbol{x}'|e^{iH(t'-t)}|\boldsymbol{x}\rangle, \quad (7.190)$$

where the Hamilton operator in the $\boldsymbol{x}$ representation is $\omega_{\boldsymbol{x}} = \sqrt{m^2 - \nabla^2}$. Using (7.181), the same localized state $f(0,\boldsymbol{x}) = \delta^3(\boldsymbol{x} - \boldsymbol{x}_0)$ can be expressed in terms of the functions $\tilde{f}(0,\boldsymbol{x})$ as

$$\tilde{f}(0,\boldsymbol{x}) = (2\omega_{\boldsymbol{x}})^{-1/2} f(t,\boldsymbol{x}) \equiv \left(2\sqrt{m^2 - \nabla^2}\right)^{-1/2} \delta^3(\boldsymbol{x} - \boldsymbol{x}_0). \quad (7.191)$$

In the case (ii), the initial wave packet (and analogously the final wave packet) is determined by (7.186), so that

$$|\Psi(0)\rangle = \int \mathrm{d}^3x\, 2\omega_{\boldsymbol{x}} \tilde{f}(0,\boldsymbol{x}) \tilde{a}^\dagger(\boldsymbol{x})|0\rangle = \tilde{a}^\dagger(\boldsymbol{x}_0)|0\rangle \equiv |\tilde{\boldsymbol{x}}_0\rangle. \quad (7.192)$$

The scalar product (7.187) can then be written in the form

$$\tilde{G}(t',\boldsymbol{x}';t,\boldsymbol{x}) = \langle \tilde{\boldsymbol{x}}'|e^{iH(t'-t)}|\tilde{\boldsymbol{x}}\rangle. \quad (7.193)$$

We have thus two kinds of propagators, (7.190) and (7.193), one between the states $|\boldsymbol{x}\rangle$, $|\boldsymbol{x}'\rangle$, and the other one between the states $|\tilde{\boldsymbol{x}}\rangle$, $|\tilde{\boldsymbol{x}}'\rangle$, which are all particular cases of a generic single particle state

$$|\Psi(0)\rangle = \int \mathrm{d}^3x\, f(0,\boldsymbol{x}) a^\dagger(\boldsymbol{x})|0\rangle = \int \mathrm{d}^3x\, 2\omega_{\boldsymbol{x}} \tilde{f}(0,\boldsymbol{x}) \tilde{a}^\dagger(\boldsymbol{x})|0\rangle \quad (7.194)$$

for two different choices, (7.184) and (7.191), of the wave packet profiles.

An explicit expression for $G(t,\boldsymbol{x}';t,\boldsymbol{x})$ is given by the expression (7.203) [that will be considered in more detail later] or the corresponding three dimensional expression considered in Ref. [175], whilst the explicit expression for the propagator (7.193) is [176–178]

$$\tilde{G}(t', \boldsymbol{x}'; t, \boldsymbol{x}) = \frac{1}{\pi^2} \frac{m^2}{\sqrt{r^2 - t^2}} K_1\left(m\sqrt{r^2 - t^2}\right), \quad r^2 = (\boldsymbol{x}' - \boldsymbol{x})^2.$$
$$(7.195)$$

From the latter expression it follows that the amplitude for the transition between the events separated by a space-like interval does not vanish. This fact has been explored within the context of the Dirac field in Ref. [178], where it was argued that contrary to the common understanding conveyed in the modern literature, such an effect may have observable macroscopic consequences.

7.3   Motion of wave packets

We will now consider solutions of the Schrödinger equation (7.42) for single particle wave packets. Using a Lorentz frame in which $n^a = (1,0,0,0)$, which gives $s = t$, $\omega_{\bar{\boldsymbol{x}}} = \omega_{\boldsymbol{x}} = \sqrt{m^2 + (-i\nabla)^2}$, Eq. (7.42) reads

$$i\frac{\partial f}{\partial t} = \omega_{\boldsymbol{x}} f. \qquad (7.196)$$

Its solution is

$$f(t, \boldsymbol{x}) = e^{-i\omega_{\boldsymbol{x}} t} f(0, \boldsymbol{x}). \qquad (7.197)$$

A single particle state is then

$$|\Psi(t)\rangle = \int d^3 \boldsymbol{x}' \, e^{-i\omega'_{\boldsymbol{x}} t} f(0, \boldsymbol{x}') a^\dagger |0\rangle. \qquad (7.198)$$

Projecting it onto $\langle \boldsymbol{x}| \equiv \langle 0|a(\boldsymbol{x})$ we obtain

$$\langle \boldsymbol{x}|\Psi(t)\rangle \equiv f(t, \boldsymbol{x}) = \int d^3 \boldsymbol{x}' \left(e^{-i\omega'_{\boldsymbol{x}} t} f(0, \boldsymbol{x}')\right) \delta^3(\boldsymbol{x} - \boldsymbol{x}'), \qquad (7.199)$$

which is equal to

$$f(t, \boldsymbol{x}) = \int d^3 \boldsymbol{x}' \, f(0, \boldsymbol{x}') \left(e^{-i\omega'_{\boldsymbol{x}} t} \delta^3(\boldsymbol{x} - \boldsymbol{x}')\right). \qquad (7.200)$$

Therefore,

$$f(t, \boldsymbol{x}) = \int d^3 \boldsymbol{x}' \, G(t, \boldsymbol{x}; 0, \boldsymbol{x}') f(0, \boldsymbol{x}'), \qquad (7.201)$$

where

$$G(t, \boldsymbol{x}; 0, \boldsymbol{x}') = e^{-i\omega'_{\boldsymbol{x}} t} \delta^3(\boldsymbol{x} - \boldsymbol{x}') \qquad (7.202)$$

is the Green function.

An explicit expression for the Green function in one dimension, valid for all values of x and t, was derived by Al-Hashimi and Wiese [165] (see also Ref. [175]):

$$G(t, x; 0, 0) \equiv G(t, x) = -\frac{imt}{\pi\sqrt{x^2 - t^2}} K_1\left(m\sqrt{x^2 - t^2}\right), \qquad (7.203)$$

where K_1 is the modified Bessesl function of degree one.

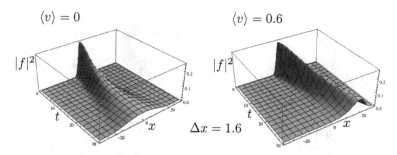

Fig. 7.1 Evolution of the probability density, $|f|^2$, for a minimal position-velocity uncertainty wave packet, $f(t, x)$, whose width is $\Delta x > \lambda_c = 1/m$, for two different velocities $\langle v \rangle$. We can express m in arbitrary units, therefore we take $m = 1$.

As shown in Ref. [165], a minimal position velocity uncertainty wave packet can be expressed as

$$f(t, x) = A G(x - i\beta, t - i\alpha). \qquad (7.204)$$

Here A is a normalization constant, and α, $\beta = \beta_R + i\beta_I$ the constants, related to the parameters of the wave packet, according to

$$\alpha = \frac{1}{2(\Delta v)^2}\langle\partial_p^2 E\rangle, \qquad \beta_R = \alpha\langle v\rangle, \qquad \beta_I = -\langle x\rangle, \qquad (7.205)$$

where $\Delta v = \sqrt{\langle v^2\rangle - \langle v\rangle^2}$.

Taking the absolute square of the wave packet (7.204), we obtain the probability density $|f(t, x)|^2$ as function of time and position. Using the computer algebra system Mathematica, it is straightforward to calculate the probability density for various choices of the parameters Δx and $\langle v\rangle$. Since the parameter β_I determines the initial position of the wave packet, we fixed it to $\beta_I = 0$. We used arbitrary units[6], therefore without loss of generality, we fixed mass to $m = 1$.

---

[6]Or we can use the extended Planck units [49] (see also Wikipedia [179]) in which $\hbar = c = G = 4\pi\epsilon_0 = 1$. In those units all physical quantities are just numbers (multiples of the fundamental Planck unit, which is the same for every quantity).

In Fig. 7.1 we show how the wave packet evolves if its initial width Δx is greater than the Compton wavelength, so that $\Delta x > \lambda_c = \frac{\hbar}{mc} = \frac{1}{m} = 1$. We see that in such case we have just the usual wave packet evolution with the probability density picked on the classical trajectory. The cases for two different velocities, namely, $\langle v \rangle = 0$ and $\langle v \rangle = 0.7$ are shown.

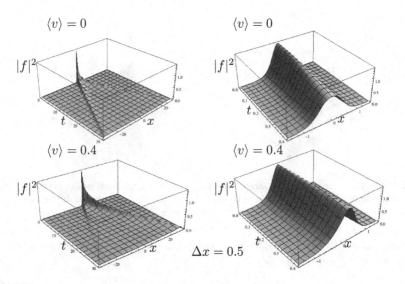

Fig. 7.2 Evolution of the probability density, $|f|^2$, for a minimal position-velocity uncertainty wave packet, $f(t, x)$, whose width is $\Delta x < 1/m$, for two different values of $\langle v \rangle$. Initially the wave packet evolves normally (lower plots), but after certain time it splits into two branches (upper plots).

In Fig. 7.2 we have examples of wave packets whose initial width is smaller than the Compton wavelength, i.e., $\Delta x < \frac{1}{m} = 1$. In this case the evolution is like the usual wave packet evolution only for a certain initial period. Afterwile, the wave packet splits into two branches, each moving on average with the speed of light, $c = 1$, into the opposite direction. The expectation value of velocity after the split remain the same. Its value is encoded in different intensities of the separated wave packets.

In Figs. 7.1 and 7.2 we considered wave packets or two different expected velocities $\langle v \rangle = 0$ and $\langle v \rangle = 0.6$ as observed from a fixed reference frame. They are related by an active Lorentz transformation. We can as well observe a wave packet, say the one with $\langle v \rangle = 0$, from two different Lorentz frames. In Ref. [166] it stays:

Inspecting the wave packets of Figs. 7.1 and 7.2, it is obvious that when observed from another Lorentz frame nothing unusual happens. In another Lorentz frame they become Lorentz transformed wave packets. If the initial width decreases, then the probability density $|f|^2$ becomes higher and higher, as shown in Fig. 7.3. In the limit of a δ-like localized wave packet at $t = 0$, $|f|^2$ becomes infinitely high and infinitely narrow, concentrated on the light cone, according to $|f|^2 = \frac{1}{2}\left(\delta(t - x) + \delta(t + x)\right)$. The event at $t = 0$ and $x = 0$ at which the particle is initially localized, is, of course, invariant in all Lorentz frames. Thus all observers see the particle localized in the origin of their Lorentz frame. At later times $t > 0$ the particle is localized on the intersection of the simultaneity hypersurface with the light cone. For such a limiting state, their is no instantaneous spreading of the probability density of the sort considered in Refs. [163, 180–182]. We thus see that the relativistic wave packet in the limit of the δ-like initial localization in fact remedies the non relativistic case, in which an infinitely thin wave packet, exactly localized at $t = 0$, spreads over all space at arbitrarily small $t > 0$.

We have also seen that the relativistic expression (7.204), derived from (7.200), (7.203) describes wave packets of any velocity, including zero velocity. Thus even a particle moving with zero velocity is described by the *relativistic wave packet*. The non relativistic wave packet is obtained from expression (7.201) in the approximation $m^2 \gg p^2$ in which we neglect higher momenta. Equivalently, it is obtained from expression (7.204) if the wave packet width Δx is large in comparison with the Compton length.

The case in which at $t = 0$ a wave is not a minimal position-velocity wave packet, but an exactly localized (rectangular) wave packet, was considered by Karpov et al. [183]. It was found that such wave packet is a superposition of two non local wave packets moving in the opposite directions with the velocity of light. Initially this gives a rectangular localized wave packet, which immediately delocalizes at $t > 0$. This is similar to the behavior of a minimal position-velocity wave packet, whose width Δx is smaller than the Compton wavelength, with the difference that the separation into two distinct wave packets becomes manifest immediately, and not after certain period. Such exact initial localization (as a rectangular wave packet), of course, is not invariant under Lorentz transformations. When observed from another frame, the simultaneity hypersurface Σ' is no longer the same, and it intersection with the evolving wave packet does not give an exact rectangular localization on

Σ', but a localization with an infinite tail. The exception, as we have seen above, is the limiting case when the width of the exact localization goes to zero and we approach the localization at a spatial point. Such, initially δ-like localized, wave packet does not instantly evolve into a wave packet with infinite tail, but remains localized on the light cone.

Concerning the usage of the word "localization", I am using it in the sense of comprising any of the four types of localization, suggested in Ref. [166]. The prevailing opinion in the literature is that they are all problematic. Namely, the three types of localization, (i) a delta-like, (ii) localization in a finite region and vanishing outside, and (iii) localization in a finite region and decaying outside, are considered as problematic. Namely, according to the Hegerfeldt also the type (iii) localization leads to causality violation. Eventually the fourth type (iv), i.e., the "effective" localization like a Gaussian wave packet, is not considered as problematic. But in opinion of numerous researchers, the wave packets of relativistic quantum mechanics, such as those plotted in the Figs. 7.1 and 7.2, cannot be consistently considered as one-particle [probability] densities. This means that also the localization of the type (iv) is often considered as problematical. I am clarifying here all four types of localization.

A state $|\boldsymbol{x}\rangle = a^\dagger(\boldsymbol{x})|0\rangle$ is an eigenstate of the position operator. A generic single particle state can be expanded in terms of such basis states according to Eq. (7.34) by means of a wave packet profile $f(t, \boldsymbol{x})$. It is the wave packet profile $f(t, \boldsymbol{x})$, or, equivalently $\tilde{f}(t, \boldsymbol{x})$, which determines how a particle is localized, not a basis state $|\boldsymbol{x}\rangle = a^\dagger(\boldsymbol{x})|0\rangle$ itself. If $f(0, \boldsymbol{x}) = \delta^3(\boldsymbol{x} - \boldsymbol{x}_0)$, then in this representation the particle is localized at $\boldsymbol{x}_0$. (What happens at later times, is determined by Eq. (7.200)–(7.204), and illustrated in Figs. 7.1 and 7.2). Between f and $\tilde{f}$ there is a nonlocal transformation, and a similar nonlocal transformation between $a^\dagger$ and $\tilde{a}^\dagger$ (see Eq. (7.37)). If at an initial time f is equal to a δ-function, then $\tilde{f}$ is not a δ-function, and vice versa. The same for the operators a and $\tilde{a}$. Because the (Newton–Wigner) position operator (7.29) is generally recognized as appropriate position operator (apart from the well-known difficulties which are clarified in this book), it is reasonable to consider its eigenstates $|\boldsymbol{x}\rangle = a^\dagger(\boldsymbol{x})|0\rangle$ as the states with definite position. The observation by Al–Hashimi and Wisse that the operator $a^\dagger(\boldsymbol{x})$ is non-local is based on their non-local relation (B.17) (which corresponds to my equation (7.37)) between $a^\dagger(\boldsymbol{x})$ and the field operators. However, a generic state is a superposition (7.34) formed either in terms of $a^\dagger(\boldsymbol{x})$ and $f(t, \boldsymbol{x})$ or $\tilde{a}^\dagger(\boldsymbol{x})$ and

$\tilde{f}(t, \boldsymbol{x})$, where $\tilde{a}(\boldsymbol{x}) = \varphi(0, \boldsymbol{x})$ is the field operator. In order to understand the issue of localization one should not only consider the basis states, which in one representation are created with $\tilde{a}^\dagger(\boldsymbol{x})$ and in another representation with $a^\dagger(\boldsymbol{x})$, but also the superposition coefficients, which in the former representation are $\tilde{f}(t, \boldsymbol{x})$ and in the latter representation are $f(t, \boldsymbol{x})$. The fact that the relation between those two representations is nonlocal does not imply that a state $|\Psi\rangle = a^\dagger(\boldsymbol{x})|0\rangle$ (with $f(0, \boldsymbol{x}') = \delta^3(\boldsymbol{x}' - \boldsymbol{x})$) is not a state with definite position. The same state in another representation is $|\Psi\rangle = \int \mathrm{d}^3\boldsymbol{x}\, \tilde{f}(0, \boldsymbol{x})\tilde{a}^\dagger(\boldsymbol{x})|0\rangle$, where $\tilde{f}$ is a function spread over all values of $\boldsymbol{x}$, and yet it is an eigenfunction of the position operator. However, the basis states $\tilde{a}^\dagger(\boldsymbol{x})|0\rangle$ are of course not eigenstates of the position operator.

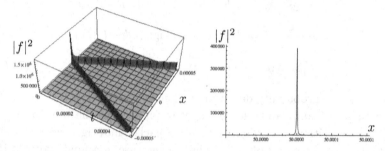

Fig. 7.3　An example of the probability density, $|f|^2$, for a wave packet, whose initial width, $\Delta x = 0.00045$, is very small in comparison with the Compton length, which in our units is $\lambda_c = 1$. We see that the probability density with decreasing Δx becomes more and more concentrated on the light cone.

In view of the above explanation, $|f|^2$ and thus the wave packets plotted in figures are indeed one particle probability densities. They behave as shown in the plots. Barat and Kimbal [184]) in their derivation did not use the function f, for which $f^*f > 0$. They used the function $\tilde{f}$ (in their notation ψ), with which one cannot define a positive probability density, but a charge density that is not always positive even for positive-frequency solutions of the Klein-Gordon equation. Therefore their result does not hold for the wave function f used in the plots of this chapter. Because the wave functions f contains only positive frequencies, a belief of the sort that the wave splitting is due to the presence of positive-energy and negative-energy contribution has no foundation.

It is true that the "superluminal spreading" of a wave packet is often related to the non-locality of the operator $\sqrt{m^2 - \nabla^2}$. But Wagner et al. (Ref. [164]) have pointed out that an infinite propagation speed is actually rather universal and not unique to the specific square root functional form of the Hamiltonian. Concerning a δ-like localized initial wave packet I have shown that there is no instantaneous spreading of the probability density of the sort considered in Refs. [163,180,181]. The probability density is concentrated on the light cone and zero elsewhere (Fig. 7.3). This is consistent with the results derived in Refs. [164, 189], and Karpov et al. (Ref. [183], where a rectangular initial wave packet is considered. Karpov et al. showed that such a momentary rectangular wave packet at $t = 0$ is formed as a superposition of two incoming non-rectangular wave packets moving with the speed of light, given by the expressions

$$
f = \frac{i}{2\sqrt{2b\pi}} \left(\log \left| \frac{b - t + x}{-b - t + x} \right| - \log \left| \frac{b + t + x}{-b + t + x} \right| \right)
$$

$$
+ \frac{1}{\sqrt{2b}} \left(\text{Sign} \left(b - t + x \right) - \text{Sign} \left(-b - t + x \right) \right)
$$

$$
+ \text{Sign} \left(b + t + x \right) - \text{Sign} \left(-b + t + x \right) \right). \tag{7.206}
$$

At $t > 0$ the probability density is no longer rectangular, but is concentrated on (and leaking outside) the light cone, because the wave packet is now a superposition of two outgoing wave packet moving in the opposite directions with the speed of light. Thus in the case of a final width of the initial wave packet (the width of the initial rectangle) there is a superluminal spreading of the probability density. In the limit when the width goes to zero, the probability density approaches a δ-distribution concentrated on the light cone. The same limiting distribution of the probability density we have obtained as a limiting case of a minimal position–velocity wave packet plotted in Fig. 7.3. In Fig. 7.4 there are some examples of the calculations based on the wave packet shape of the above references [164, 189], and [183].

We have considered the motion of wave packets in the absence of interactions. But a big virtue of quantum field theory is its ability to consider multi particle states and interactions among them A free single particle state is an idealisation. There is no such states in the universe. All states are mutually connected, entangled and interacting in general. And yet in non relativistic quantum mechanics it is possible to consider interaction free wave packets, and even realize them experimentally to a high degree of accuracy. We have shown that free single particle wave packet states can also be consistently described within the framework of relativistic quantum field theory. By clarifying the issues of such states, we have clarified the concept

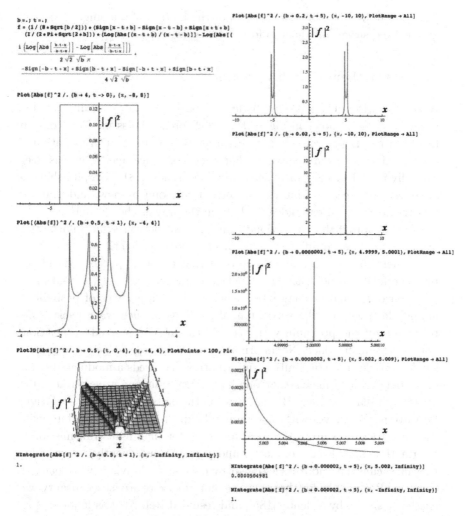

Fig. 7.4 Evolution of an initially rectangular wave packet as considered by Karpov et al. The same wave packet has been considered by Eckstein and Miller, and Wagner et al. In the limit when the packet width goes to zero, the probability density approaches to a δ-like distribution concentrated on the light cone.

of position and its role in quantum field theory. So we have strengthened the theoretical and conceptual foundations of quantum field theory, and made it clearer and easier to understand intuitively. A prerequisite for this is understanding and describing interaction free single particle states. Analogously, the transition from Aristotelian to Galilean and Newtonian physics required clear understanding of the motion of a a single object in an idealized situation when no forces are acting on it.

7.4   What about causality violation?

A common wisdom is that relativistic wave packets violate causality. In fact, the wave packets shown in Figs. 6.1 and 6.2 spread across the light cone. It has been taken for granted that this implies superluminal propagation of causal influences from one to another region of spacetime. This has been carefully and rigorously investigated by Hegerfeldt [181]. Despite that his analysis is impeccable from the mathematical point of view, and that the relativistic wave packets indeed behave in the way as shown by Hegerfeldt, this is not a proof that they violate causality in the sense that information or any sort of causal influence can travel faster than light.

Namely, one has to take into account that the wave packets considered by Hegerfeldt are not classical bunches of matter that exert causal influences on other matter they interact with. They are waves of probability amplitude, therefore the superluminal portions of such wave packets determine only the probability that particles will be detected at positions outside the light cone. No information transfer can be achieved with a single particle. For transmitting information one needs a modulated beam of particles, which means that one has to transmit one wave packet after another, or many at once. But as shown in the preceding section, a relativistic wave packet moves with a subluminal expected speed. Therefore, with such particles on average there can be no faster than light transmission of information, and also not of matter and energy. Consequently, no "grandfather like" causal paradoxes can occur by means of relativistic wave packets. That the Reeh-Schlieder theorem does not violate relativistic causality was explicitly shown by Valente [185], and resumed Ref. [166] as follows:

> We conclude that the usual arguments against localized relativistic states can be circumvented. Such states naturally occur within quantum field theory and are not problematic at all. This sheds new light on the implications of the Reeh-Schlieder theorem [187], which is interpreted as implying that states (including single particle states) cannot be exactly localized in a finite region (see, e.g., [165]). Such a conclusion comes from the fact that one of the axioms of algebraic quantum field theory [186] is causality.

However, as pointed out by G. Valente [185], one has to distinguish among different concepts of 'causality' used in the literature, and not all imply the possibility of information transmission. Moreover, Karpov et al. [183] and Antoniou et al. [188] have demonstrated that the classical measurement cannot detect the "acausal" effects of the wave packet quantum states. In the scenario that occurred in the Reeh-Schlieder theorem, the superluminal influence of a field in one spacetime region to a field in another region cannot be used for a controlled transmission of information. Therefore, the Reeh-Schlieder theorem does not imply that quantum states cannot be localized in a finite region. They can be localized, but their immediate spreading over all the space, cannot be used for a superluminal transmission of information.

That the issues concerning causality and relativistic wave packet localization is not as straightforward and clear as usually assumed has been also pointed out by Fleming [190, 191] and Wagner [164]. Moreover, Ruijsenaars [192] has found that because of the smallness of the effect the detection of the so called acausal events is not possible with the present technology. Anyway, the transmission of information into the past or future does not necessarily lead to causal paradoxes. One only needs to step out of the box that so far has been constraining the conceptual foundations of physics. In the next section we will discuss tachyons and show at the end that they imply no problems with causality, provided that one takes into account a proper interpretation of quantum mechanics.

Chapter 8

Misconceptions and Confusion About Tachyons

We have seen how within the framework of relativistic quantum field theory the wave packet profiles of single particle states represented in position space make perfect sense, in spite of the prevailing opinion that such states lead to inconsistencies, and thus make no sense. Besides the confusion regarding the Lorentz invariance and covariance of such states, the main argument against relativistic localized states was that they violate causality, because they leak outside the light cone. Our calculations show that such leakage is very small, and cannot be used for superluminal transmission of information, and therefore no causality violation in the sense of the grandfather paradox can take place. In this chapter we will go a step further and consider the case in which superluminal signals are indeed theoretically possible by means of tachyons, the particles that travel faster than light. We are going to report about the findings of Ref. [109].

8.1 Introduction

Today, tachyons [194] are considered as impossible particles. A typical argument is that they violate causality. This was nicely illustrated by the setup of the "tachyonic antitelephone" [195], according to which tachyons, considered as localized particles propagating faster than light, can transmit information into the past, or bring it from the future. There are good reasons to consider this as paradoxical and hence conclude that superluminally propagating tachyons cannot exist. Such conclusion is confirmed by the investigations of the tachyonics fields, satisfying, e.g., the Klein-Gordon equation with the negative squared mass [196]. The initial data for such a field cannot be arbitrarily specified. Therefore, we cannot form an initial wave packet and propagate it with a superluminal velocity.

On the other hand, the so called *extended relativity* has been investigated that considers not only *the subluminal*, but also *superluminal Lorentz transformations* (SLT) [197–199]. Those transformations change the sign of the quadratic form, namely, $x'^\mu x'_\mu = -x^\mu x_\mu$, and $p'^\mu p'_\mu = -p^\mu p_\mu$, $\mu = 0, 1, 2, 3$. Under a superluminal boost in the x^1-direction, the real coordinates x^2, x^3 and the momenta p^2, p^3 become imaginary. The Klein-Gordon equation in the new reference frame has the same form as in the original frame, but with x^0 and x^1 interchanged. The "initial data" have now to be speficied on a time-like hypersurface $x^1 = constant$. For the group velocity of the field we keep the definition dp^0/dp^1, which is greater than the velocity of light.

According to the extended relativity, a tachyonic field is the one obtained by a bradyonic field by a SLT. A field satisfying the Klein-Gordon equation with the negative mass square is not the tachyonic field of the extended relativity, because it cannot be obtained from a bradyonic field by a SLT; it is a completely different object. We will show that the tachyonic field of extended relativity, i.e., the field obtained by a bradyonic field by a SLT, propagates with a superluminal group velocity[1].

But superluminal transformations cannot be realized in real spacetime $M_{1,3}$, they need complex spacetime $M_{1,3} \times \mathbb{C}$, which we can replace with a real eight-dimensional space $M_{4,4}$ of signature $(4, 4)$. In such space, the Cauchy problem is not well posed, unless the hypersurface is light-like. On a generic (7-dimensional) hypersurface, e.g., $x^0 = constant$, we cannot freely specify initial data and then calculate the behaviour of the field at later times. But it turns out that if we have in advance certain information about how the field behaves in the transverse directions, e.g., in the subspace $M_{3,1}$, then we can freely specify the initial data on a space-like 4-dimensional surface of the subspace $M_{1,3}$ and calculate the subsequent evolution of the field at later times.

A particular example of a higher dimensional space with important physical consequences is the 16-dimensional Clifford space C [49, 70, 71, 99, 104, 106]. It is a manifold that at any of its points has the Clifford algebra $Cl(1, 3)$ of spacetime $M_{1,3}$ as tangent space. Mathematically, C is the space of oriented r-volumes, $r = 0, 1, 2, 3$, that can be associated with physical extended objects living in 4D spacetime. spacetime. The signature of C is $(8, 8)$. The Cauchy data can be specified on a 14D surface with

---

[1]The usual concepts such as the tachyon localization, condensations due to the instability in the presence of negative energies, and the Čerenkov radiation do not hold for the tachyons of extended relativity.

signature $(7,7)$. The remaining two dimensions, of which one is time-like and one space-like, can be combined so to serve as the Stueckelberg evolution parameter τ. In Clifford space we can thus formulate the Stueckelberg theory [49,112–122] in which tachyons are consistent objects.

Finally we point out that the notorious causal paradoxes of tachyons can be resolved if one does not consider tachyons as classical objects tracing deterministic paths, but as the quantum objects with positions spread à la wave packets.

8.2 The superluminal transformations of extended relativity

According to the research performed in Refs. [198,199], special relativity is not the last word about the dynamics of objects in classical physics. Relativity can be extended so to incorporate slower and faster than light particles, *bradyons B* and tachyons T, that can be transformed into each other by *superluminal transformations* (SLT). If in a reference frame S a particle is observed as bradyon (having a velocity $v < c$), then in a superluminal reference frame S' the same particle is observed as a tachyon (having $v > c$). While subluminal Lorentz transformations preserve the quadratic form, $\mathrm{d}s^2 = \eta_{\mu\nu}\mathrm{d}x^\mu\mathrm{d}x^\nu$, so that $\mathrm{d}s'^2 = \mathrm{d}s^2$, the superluminal Lorentz transformations change the sign, so that $\mathrm{d}s'^2 = -\mathrm{d}s^2$.

An example is the superluminal transformation (a boost) in the x-direction:

$$t' = \frac{t + vx}{\sqrt{v^2 - 1}} \,, \qquad x' = \frac{vt + x}{\sqrt{v^2 - 1}} \,, \qquad y' = iy \,, \qquad z' = iz. \qquad (8.1)$$

The transformation of y and z must involve the imaginary unit i, because only then the quadratic form changes the sign, i.e.,

$$\mathrm{d}t'^2 - \mathrm{d}x'^2 - \mathrm{d}y'^2 - \mathrm{d}z'^2 = -\left(\mathrm{d}t^2 - \mathrm{d}x^2 - \mathrm{d}y^2 - \mathrm{d}z^2\right). \qquad (8.2)$$

The analogous transformations hold for the 4-momentum $p^\mu = (p^0, p^1, p^2, p^3) \equiv (p_t, p_x, p_y, p_z)$, where $p^0 \equiv p_t \equiv E$ is a particle's energy. A SLT changes $p^\mu p_\mu$ into $-p^\mu p_\mu$ so that

$$p'^\mu p'_\mu = -p^\mu p_\mu. \qquad (8.3)$$

A bradyon in S' satisfies

$$p'^\mu p'_\mu = m^2, \qquad (8.4)$$

which under a SLT becomes

$$-p^\mu p_\mu = m^2. \qquad (8.5)$$

The mass m is assumed to be invariant under SLT.

Writing Eqs. (8.4) and (8.5) explicitly, we have that a bradyon in S' satisfies

$$(p'^0)^2 - (p'^1)^2 - (p'^2)^2 - (p'^3)^2 = m^2, \tag{8.6}$$

where p'^0, p'^1, p'^2, p'^3 are all real, while the same particle observed in S is a tachyon, satisfying

$$-(p^0)^2 + (p^1)^2 + (p^2)^2 + (p^3)^2 = m^2. \tag{8.7}$$

If S and S' are related by the superluminal transformation (8.1), then p^0, p^1 are *real*, while p^2 and p^3 are *imaginary*. Rewriting (8.7) in terms of the real quantities $\tilde{p}^2 = ip^2$, $\tilde{p}^3 = ip^3$, Eq. (8.7) becomes

$$(p^1)^2 - (p^0)^2 - (\tilde{p}^2)^2 - (\tilde{p}^3)^2 = m^2. \tag{8.8}$$

We see that this is just the usual, bradyonic, mass shell constraint with p^1 and p^0 interchanged.

Upon quantization, we replace the momenta with the operators $p'_\mu = -i\partial'_\mu$, $\partial'_\mu \equiv \partial/\partial x'^\mu$, and the constraint (8.6) becomes the Klein-Gordon equation

$$\left(-\partial'_\mu \partial'^\mu - m^2\right)\phi'(x') = 0, \tag{8.9}$$

its particular solution being

$$\phi'(x') = e^{ip'_\mu x'^\mu} = e^{i(p'_t - p'_x x' - p'_y y' - p'_z z')}. \tag{8.10}$$

This holds in the frame S'.

In the frame S, obtained from S' by the SLT given in Eq. (8.1), we have the Klein-Gordon equation corresponding to the constraint (8.7), which is equivalent to the constraint (8.8):

$$\left(\partial_\mu \partial^\mu - m^2\right) = \left(\frac{\partial^2}{\partial t^2} - \frac{\partial^2}{\partial x^2} - \frac{\partial^2}{\partial y^2} - \frac{\partial^2}{\partial z^2} - m^2\right)\phi(t, x, y, z)$$

$$= \left(\frac{\partial^2}{\partial t^2} - \frac{\partial^2}{\partial x^2} + \frac{\partial^2}{\partial \tilde{y}^2} + \frac{\partial^2}{\partial \tilde{z}^2} - m^2\right)\phi(t, x, \tilde{y}, \tilde{z})$$

$$= 0. \tag{8.11}$$

Mathematically, the latter equation has the same form as the bradyonic Klein-Gordon equation in which t and x are interchanged.

A particular solution of Eq. (8.11) is

$$\phi(t, x, \tilde{y}, \tilde{z}) = e^{-i(p_t t - p_x x - p_y y - p_z z)}$$

$$= e^{-i(p_t t - p_x x + \tilde{p}_y \tilde{y} - \tilde{p}_z \tilde{z})}, \tag{8.12}$$

where the imainary quantities p_y, p_z, y, z are expressed in terms of the real quantities $\tilde{p}_y$, $\tilde{p}_z$, $\tilde{y}$, $\tilde{z}$ according to

$$\tilde{p}_y = ip_y , \quad \tilde{p}_z = ip_z , \quad \tilde{y} = iy , \quad \tilde{z} = iz. \tag{8.13}$$

A general solution of Eq. (8.11) is

$$\phi(t, x, \tilde{y}, \tilde{z}) = \int dp_t dp_x d\tilde{p}_y d\tilde{p}_z c(p_t, p_x, \tilde{p}_y, \tilde{p}_z) e^{i(p_x x - p_t t - \tilde{p}_y \tilde{y} - \tilde{p}_z \tilde{z})}$$

$$\times \delta(p_x^2 - p_t^2 - \tilde{p}_y^2 - \tilde{p}_z^2 - m^2). \tag{8.14}$$

The superposition coefficients $c(p_t, p_x, \tilde{p}_y, \tilde{p}_z)$ are functions of momenta, restrictred to the mass shell $p_x^2 - p_t^2 - \tilde{p}_y^2 - \tilde{p}_z^2 - m^2 = 0$, imposed in the integrand by the δ-function.

We now introduce the quantity

$$\omega_x = |\sqrt{m^2 + p_t^2 + \tilde{p}_y^2 + \tilde{p}_z^2}| \tag{8.15}$$

and integrate over $p_x = \pm \omega_x$. The result is an expression, analogous to the one for the bradyonic scalar field:

$$\phi = \int dp_t d\tilde{p}_y d\tilde{p}_z \frac{1}{2\omega_x} \left[e^{i(\omega_x x - p_t t - \tilde{p}_y \tilde{y} - \tilde{p}_z \tilde{z})} c(\omega_x, p_t, \tilde{p}_y, \tilde{p}_z) \right.$$

$$\left. + e^{i(-\omega_x x - p_t t - \tilde{p}_y \tilde{y} - \tilde{p}_z \tilde{z})} c(-\omega_x, p_t, \tilde{p}_y, \tilde{p}_z) \right], \tag{8.16}$$

in which the role of ω_p and ω_x (and of t and x) is interchanged. Therefore, the initial data now cannot be given on a hypersurface $x^0 \equiv t = constant$. Instead, they are given on a hypersurface $x^1 \equiv x = constant$, which is a time-like 3-surface spanned by the coordinates $(t, \tilde{y}, \tilde{z})$. Once the initial data on a hypersurface $x = constant$ are known, one can calculate the field ϕ on other hypersurface with different values of x. In the literature has already been discussed that a Cauchy surface for space-like states must be time-like: in Refs. [200, 201] without employing SLT, and in Ref. [202, 203] within the formalism of SLT in 2-dimensional spacetime.

Let us now investigate the motion of wave packets satisfying the bradyonic Klein-Gordon equation (8.9) (in the frame S') and its superluminal transform (8.11). In Chapter 7 we showed that though a field $\phi(x)$ satisfying the Klein-Gordon equation cannot be interpreted as the probability amplitude for observing the particle at position x, the combination $\psi = \phi + \omega_x^{-1} \dot{\phi}$ has the role of a wave function. It satisfies Eq. (7.204). The probability

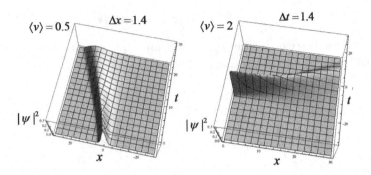

Fig. 8.1   An example of a bradyonic (left), and a tachyonic (right) wave packet.

density $|\psi|^2$ as function of t and x is plotted in Fig. 7.1. For a tachyonic wave function we have mathematically the same equation, but with t and x interchanged. In Fig. 8.1 are shown examples of $|\psi|^2$ for a subluminal and the corresponding superluminal wave packet.

In Ref. [109] the wave packet for non-relativistic approximation to the Klain-Gordon equation were considered. The obtained plots were similar to those in Fig. 8.1. The probability density in the frame S' is in this approximation given by a Gaussian wave packet

$$|\psi'|^2 \propto \exp\left[-\frac{(\mathbf{x}' - \frac{\mathbf{P}'_0}{m}t')^2}{\sigma'(t')}\right], \qquad (8.17)$$

where $\sigma'(t') = \sigma'_0 + t'^2/(m^2\sigma'_0)$. The probability density is picked on the classical trajectory $\mathbf{x}' = \frac{\mathbf{P}'_0}{m}t$ with velocity $\mathbf{v}' = \frac{\mathbf{P}'_0}{m}$, $|\mathbf{v}'| < c = 1$. The Klein-Gordon equation (8.9) in the reference frame S' thus describes a field propagating with a slower than light velocity $\mathbf{v}'$ (Fig. 8.1 left).

The probability density in the frame S, related to S' by a SLT is given by the superluminally transformed $|\psi'|^2$, given by

$$|\psi|^2 \propto \exp\left[-\frac{(t - \frac{p_{t0}}{m}x)^2}{\sigma(x)} - \frac{(\tilde{y} - \frac{\tilde{p}_{y0}}{m}x)^2}{\sigma(x)} - \frac{(\tilde{z} - \frac{p_{\tilde{z}0}}{m}x)^2}{\sigma(x)}\right], \qquad (8.18)$$

where $\sigma(x) = \sigma_0 + \frac{x^2}{m^2\sigma_0}$. The probability density is now picked on the classical trajectory, $t = \frac{p_{t0}}{m}x$, $\tilde{y} - \frac{\tilde{p}_{y0}}{m}$, $\tilde{z} - \frac{\tilde{p}_{z0}}{m}$, where

$$\frac{p_{t0}}{m} = \frac{dt}{dx}, \quad \frac{\tilde{p}_{y0}}{m} = \frac{d\tilde{y}}{dx}, \quad \frac{\tilde{p}_{z0}}{m} = \frac{d\tilde{z}}{dx}, \qquad (8.19)$$

$$\left|\sqrt{\left(\frac{dt}{dx}\right)^2 + \left(\frac{d\tilde{y}}{dx}\right)^2 + \left(\frac{d\tilde{z}}{dx}\right)^2}\right| < c = 1. \qquad (8.20)$$

If we keep on assigning to the coordinate t the role of time, then we must define the velocity of the wave packet in the x-direction as the derivative $v_x = \frac{dx}{dt} = \frac{m}{p_{t0}} > 0$. The velocity is thus superluminal.

The reciprocity between time and space when bradyons are replaced by tachyons was noticed long time ago [198,199,204]. Localization of tachyons on the basis of such reciprocity was considered by Vyšin [202, 203], Shay [200], and others [205–207].

In the full relativistic treatment we have for a tachyonic field the dispersion relation (8.8) that we will now write as

$$p_x^2 - p_t^2 - \tilde{p}_y^2 - \tilde{p}_z^2 = m^2. \tag{8.21}$$

If, using (8.21), we write the dispersion relation as

$$p_x = \sqrt{p_t^2 + \tilde{p}_y^2 + \tilde{p}_z^2 + m^2}, \tag{8.22}$$

then we have

$$\frac{dp_x}{dp_t} = \frac{dt}{dx} = \frac{p_t}{\sqrt{p_t^2 + \tilde{p}_y^2 + \tilde{p}_z^2 + m^2}} \;, \qquad \left|\frac{dt}{dx}\right| < c = 1. \tag{8.23}$$

This is the reciprocal group velocity of a tachyonic field in the x-direction. Alternatively, instead of (8.22) we can use the dispersion relation

$$p_t = \sqrt{p_x^2 - \tilde{p}_y^2 - \tilde{p}_z^2}, \tag{8.24}$$

which gives the superluminal group velocity

$$\frac{dp_t}{dp_x} = \frac{dx}{dt} = \frac{p_x}{\sqrt{p_x^2 - \tilde{p}_y^2 - \tilde{p}_z^2 - m^2}} \;, \qquad \left|\frac{dx}{dt}\right| > c = 1. \tag{8.25}$$

From the calculated wave packets (Fig. 8.1) we see that while the wave packet on the left is subluminal and loicalized in space, the wave packet on the right is superluminal and localized in time (more precisely, in the time-like 3-surface $(t, \tilde{y}, \tilde{z})$). Because the subluminal wave packet can transmit information[2], so can the superluminal wave packet, which is nothing but a superluminal transform of a subluminal wave packet.

From Fig. 8.1 we see that the spatial localization width for a tachyon is wide and can be infinite. In Ref. [109] we then read:

"The fact that tachyons are infinitely extended in space, was taken as an argument against the possibility that they can transmit information. However, as already observed by Vyšín [202, 203], it is important that

---

[2]This is a short hand way of talking. More precisely, we should say that a modulated train of wave packets can transmit information.

tachyons form sharp pulses in time. A sequence of such pulses, localized in time, can encode information. If there is an interaction between tachyons and a bradyonic detector, then a bradyonic observer B would be able to observe that sequence of pulses, and interpret it as the information, emitted by another observer A. This is illustrated in Fig. 8.2.

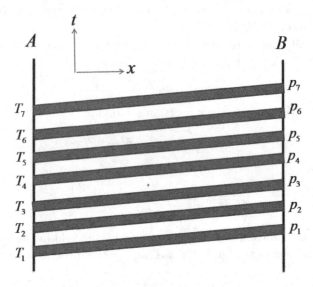

Fig. 8.2 Though the tachyon wave packets are not localized (have no sharp peaks and leading edges) in the x-direction, they can nevertheless transmit information from A to be B, because they are localized in time. Namely, if tachyons can interact with bradyons, then the observer A can send a message encoded in a sequence of emitted tachyons, T_1, T_2, T_3, ..., that are detected by the observer B as a sequence of pulses, p_1, p_2, p_3, ..., at a fixed spatial position. B can unambiguously interpret such sequence as a message, if both observers had already agreed about the code.

The solution (8.14) is the superluminal transform of the general solution of the Klein-Gordon equation (8.9) for a bradyonic field in S'. Under a SLT, the bradyonic Klein-Gordon equation transforms into the tachyonic Klein-Gordon equation. But according to the principle of relativity, the laws of motion, encrypted in the equations of motion, should remain unchanged under all transformations that bring one dynamically possible solution into another dynamically possible solution. According to our assumption, SLT are such transformations. Therefore, the equations of motion should remain invariant under SLT. Since Eq. (8.9) is not invariant, it means that it is not a complete equation, but a part of a more general equation. In the following we will consider a more general theory which the Klein-Gordon equation (8.9) is embedded in. First, we will discuss a generalization of the classical theory of the relativistic point particle, and then its quantization."

This is what we will treat in the next section.

8.3 Formulating extended relativity in real spacetime $M_{4,4}$

We have seen that extending relativity to superluminal frames by means of superluminal transformations brings imaginary coordinates and momenta into the description. Performing further boosts and rotations in different directions will lead to complex coordinates and momenta:

$$X^\mu = x^\mu + i\tilde{x}^\mu \,, \qquad P^\mu = p^\mu + i\tilde{p}^\mu. \tag{8.26}$$

The quantities $\tilde{x}^2$, $\tilde{x}^3$ and $\tilde{p}^2$, $\tilde{p}^3$ (written also as $\tilde{y}$, $\tilde{z}$ and $\tilde{p}_y$, $\tilde{p}_z$), introduced in the previous section, are part of the general scheme (8.26), which besides the usual x^μ involves the extra degrees of fredom, $\tilde{x}^\mu$, $\mu = 0, 1, 2, 3$.

Instead of working with complex quantities (8.26), we can use real quantities $(x^\mu, \tilde{x}^\mu)$ and $(p^\mu, \tilde{p}^\mu)$. Thus, instead of the 4-dimensional complex spacetime $M_{1,3} \times \mathbb{C}$, we can consider the 8-dimensional real space $M_{4,4}$ with signature $(4, 4)$. Mathematically, this is an ultrahyperbolic space with neutral signature. In the literature superluminal transformations are usually considered in 6D spacetime $M_{3,3}$ (see [208–211]).

Under a *subluminal* Lorentz transformation the quadratic form

$$ds^2 = dx^\mu dx_\mu - d\tilde{x}^\mu d\tilde{x}_\mu \tag{8.27}$$

is preserved. Under a *superluminal* Lorentz tranformation it changes sign:

$$ds'^2 = dx'^\mu dx'_\mu - d\tilde{x}'^\mu d\tilde{x}'_\mu = -ds^2 = d\tilde{x}^\mu d\tilde{x}_\mu - dx^\mu dx_\mu, \tag{8.28}$$

which means that the roles of x^μ and $\tilde{x}^\mu$ are interchanged. Thus, under a SLT the time-like coordinates $(x^0, \tilde{x}^1, \tilde{x}^2, \tilde{x}^3)$ are changed into the space-like coordinates $(\tilde{x}^0, x^1, x^2, x^3)$, and vice versa.

The same holds for the momentum quadratic form $p^\mu p_\mu - \tilde{p}^\mu \tilde{p}_\mu$. So we have that $p^\mu p_\mu - \tilde{p}^\mu \tilde{p}_\mu$ is preserved uner subluminal Lorentz transformations and changes the sign under SLT:

$$p'^\mu p'_\mu - \tilde{p}'^\mu \tilde{p}'_\mu = -(p^\mu p_\mu - \tilde{p}^\mu \tilde{p}_\mu) = \tilde{p}^\mu \tilde{p}_\mu - p^\mu p_\mu. \tag{8.29}$$

For a superluminal boost in the x^1-direction with the velocity v we have

$$\tilde{t}' = \frac{t + vx}{\sqrt{v^2 - 1}} \,, \qquad x' = \frac{vt + x}{\sqrt{v^2 - 1}} \,, \qquad y' = -\tilde{y} \,, \qquad z' = -\tilde{z}, \tag{8.30}$$

where $(x^0, x^1, x^2, x^3) \equiv (t, x, y, z)$ and $(\tilde{x}^0, \tilde{x}^1, \tilde{x}^2, \tilde{x}^3) \equiv (\tilde{t}, \tilde{x}, \tilde{y}, \tilde{z})$.

Let the momentum constraint in the reference frame S' be

$$p'^{\mu} p'_{\mu} - \tilde{p}'^{\mu} \tilde{p}'_{\mu} + M^2 = 0, \qquad (8.31)$$

where M^2 is an invariant constant, which can be either positive or negative:

$$M^2 > 0 , \qquad \text{or} \quad M^2 < 0. \qquad (8.32)$$

In a reference frame S, related to S' by a SLT, e.g., the boost (8.29), the constraint (8.31) becomes

$$-p^{\mu} p_{\mu} + \tilde{p}^{\mu} \tilde{p}_{\mu} + M^2 = 0. \qquad (8.33)$$

In the subspace $M_{1,3}$, spanned by x^{μ}, the momentum quadratic form is

$$p^{\mu} p_{\mu} = M^2 + \tilde{p}^{\mu} \tilde{p}_{\mu}. \qquad (8.34)$$

It can be time-like, $p^{\mu} p_{\mu} > 0$, space-like, $p^{\mu} p_{\mu} < 0$, or light-like, $p^{\mu} p_{\mu} = 0$, depending on the values of the quantities M^2 and $\tilde{p}^{\mu} \tilde{p}_{\mu}$, i.e., whether $\tilde{p}^{\mu} \tilde{p}_{\mu} > -M^2$, $\tilde{p}^{\mu} \tilde{p}_{\mu} < -M^2$, $\tilde{p}^{\mu} \tilde{p}_{\mu} = -M^2$. The analogous holds for the quadratic form in the subspace $M_{3,1}$ spanned by $\tilde{x}^{\mu}$.

The type of a particle, bradyon or tachyon, is determined with respect to the chosen subspace, $M_{1,3}$ or $M_{3,1}$. Under a SLT, the quadratic form (8.33) changes into (8.31), and thus a bradyon changes into a tachyon, and vice versa. Thus, if in the frame S a particle is a bradyon with respect to $M_{1,3}$, then the same particle is in S' observed as a tachyon.

In the quantized theory the constraint (8.33) becomes the Klein-Gordon equation

$$(\partial^{\mu} \partial_{\mu} - \tilde{\partial}^{\mu} \tilde{\partial}_{\mu} + M^2) \phi(x^{\mu}, \tilde{x}^{\mu}) = 0, \qquad (8.35)$$

whose general solution is

$$\phi(x^{\mu}, \tilde{x}^{\mu}) = \int \mathrm{d}^4 p \, \mathrm{d}^4 \tilde{p} \, c(p, \tilde{p}) e^{i(p_{\mu} x^{\mu} - \tilde{p}_{\mu} \tilde{x}^{\mu})} \delta(p^{\mu} p_{\mu} - \tilde{p}^{\mu} \tilde{p}_{\mu} - M^2). \quad (8.36)$$

Choosing a time-like component, say p^0, and integrating it out, the above solution reads

$$\phi(x^{\mu}, \tilde{x}^{\mu}) = \int \mathrm{d}^3 \boldsymbol{p} \, \mathrm{d}\tilde{p}^0 \, \mathrm{d}^3 \tilde{\boldsymbol{p}} \, \frac{1}{2\omega} \left[e^{i(\omega x^0 - \boldsymbol{p}\boldsymbol{x} - \tilde{p}_0 \tilde{x}^0 + \tilde{\boldsymbol{p}}\tilde{\boldsymbol{x}})} c(\omega, \boldsymbol{p}, \tilde{p}^0, \tilde{\boldsymbol{p}}) \right.$$

$$\left. + e^{i(-\omega x^0 - \boldsymbol{p}\boldsymbol{x} - \tilde{p}_0 \tilde{x}^0 + \tilde{\boldsymbol{p}}\tilde{\boldsymbol{x}})} c(-\omega, \boldsymbol{p}, \tilde{p}^0, \tilde{\boldsymbol{p}}) \right], \qquad (8.37)$$

where

$$p^0 = \pm \omega , \qquad \omega = |\sqrt{(\tilde{p}^0)^2 + \boldsymbol{p}^2 - \tilde{\boldsymbol{p}}^2 + M^2}|. \qquad (8.38)$$

Because ω is not real for all values of $\tilde{p}^0$, $\boldsymbol{p}$, $\tilde{\boldsymbol{p}}$, the initial values of the field and its derivative on a 7D hypersurface, e.g., $x^0 = 0$, cannot be arbitrarily specified. In other words, for a chosen $\phi(0, \boldsymbol{x}, \tilde{x}^\mu)$ and $\dot{\phi}(0, \boldsymbol{x}, \tilde{x}^\mu)$ it is not possible to obtain by the Fourier transform the expansion coefficients $c(\omega, \boldsymbol{p}, \tilde{p}^0, \tilde{\boldsymbol{p}})$, $c(-\omega, \boldsymbol{p}, \tilde{p}^0, \tilde{\boldsymbol{p}})$, and then the field values at times $x^0 \neq 0$, outsided the initial 7D hypersurface $x^0 = 0$. The Cauchy problem is not well posed [212].

However, for the known expansion coefficients $c(\omega, \boldsymbol{p}, \tilde{p}^0, \tilde{\boldsymbol{p}})$ and $c(-\omega, \boldsymbol{p}, \tilde{p}^0, \tilde{\boldsymbol{p}})$ the field is determined at all point of the space $M_{4,4}$. Let us now suppose that the coefficients are not given completely, but only partially, for instance:

$$c(\omega, \boldsymbol{p}, \tilde{p}^0, \tilde{\boldsymbol{p}}) = \delta^3(\tilde{\boldsymbol{p}} - \tilde{\boldsymbol{p}}_c)a(\omega, \boldsymbol{p}, \tilde{p}^0),$$
$$c(-\omega, \boldsymbol{p}, \tilde{p}^0, \tilde{\boldsymbol{p}}) = \delta^3(\tilde{\boldsymbol{p}} - \tilde{\boldsymbol{p}}_c)a(-\omega, \boldsymbol{p}, \tilde{p}^0). \tag{8.39}$$

If we choose $M^2 > 0$ and restrict $\tilde{p}_c$ according to

$$\tilde{p}_c^2 \leq M^2, \tag{8.40}$$

then ω, defined in Eq. (8.38) is real for all values of $\tilde{p}^0$, $\boldsymbol{p}$ and $\tilde{\boldsymbol{p}}$.

Inserting (8.39) into the field expansion (8.37), we obtain

$$\phi(x^\mu, \tilde{x}^\mu) = \int d^3\boldsymbol{p}\, d\tilde{p}^0 \frac{1}{2\omega_c} \left[e^{i\omega_c x^0} a(\omega_c, \tilde{p}^0, \boldsymbol{p}) \right.$$

$$\left. + e^{-i\omega_c x^0} a(-\omega_c, \tilde{p}^0, \boldsymbol{p}) \right] e^{-i(\boldsymbol{px} + \tilde{p}_0 \tilde{x}^0 + \tilde{\boldsymbol{p}}_c \tilde{\boldsymbol{x}})}, \tag{8.41}$$

where

$$\omega_c = |\sqrt{(\tilde{p}^0)^2 + \boldsymbol{p}^2 + M^2 - \tilde{\boldsymbol{p}}^2}|. \tag{8.42}$$

On a space-like 4-surface, defined by $x^0 = 0$, $\tilde{x} = 0$, the values of the field and its time derivative are

$$\phi(0, \boldsymbol{x}, \tilde{x}^0, \boldsymbol{0}) = \int d^3\boldsymbol{p}\, d\tilde{p}^0 \frac{1}{2\omega_c} \left[a(\omega_c, \boldsymbol{p}, \tilde{p}^0) + a(-\omega_c, \boldsymbol{p}, \tilde{p}^0) \right] e^{-i(\boldsymbol{pb} + \tilde{p}^0 \tilde{x}^0)}, \tag{8.43}$$

$$\dot{\phi}(0, \boldsymbol{x}, \tilde{x}^0, \boldsymbol{0}) = \int d^3\boldsymbol{p}\, d\tilde{p}^0 \frac{1}{2} \left[a(\omega_c, \boldsymbol{p}, \tilde{p}^0) - a(-\omega_c, \boldsymbol{p}, \tilde{p}^0) \right] e^{-i(\boldsymbol{px} + \tilde{p}^0 \tilde{x}^0)}. \tag{8.44}$$

Using Eqs. (8.43), (8.44), the coefficients $a(\omega_c, \boldsymbol{p}, \tilde{p}^0)$ and $a(-\omega_c, \boldsymbol{p}, \tilde{p}^0)$ can be calculated by the Fourier transformation of $\phi(0, \boldsymbol{x}, \tilde{x}^0, \boldsymbol{0})$ and $\dot{\phi}(0, \boldsymbol{x}, \tilde{x}^0, \boldsymbol{0})$. With the known $a(\omega_c, \boldsymbol{p}, \tilde{p}^0)$ and $a(-\omega_c, \boldsymbol{p}, \tilde{p}^0)$, the field $\phi(x^\mu, \tilde{x}^\mu)$ at any point of $M_{4,4}$ is determined by Eq. (8.41).

In this setup the field propagates as a plane wave $e^{-i\tilde{\boldsymbol{p}}_c\tilde{\boldsymbol{x}}}$ with a fixed $\tilde{p}_c$ along a time-like direction $\tilde{\boldsymbol{x}}$. The field is thus not localized within the 3D surface spanned by the coordinates $\tilde{\boldsymbol{x}} \equiv (\tilde{x}^1, \tilde{x}^2, \tilde{x}^3)$. Because of the factors $e^{i\omega_c x^0}$ and $e^{-i\omega_c x^0}$, it also is not localized in x^0. Altogether, the field (8.41) is not localized within the surface $(x^0, \tilde{x}^1, \tilde{x}^2, \tilde{x}^3)$. However, it can be arbitrarily shaped or localized (e.g., like a Gaussian wave packet) within a given initial 4-surface $(x^1, x^2, x^3, \tilde{x}^0)$, defined by $x^0 = 0$, $\tilde{\boldsymbol{x}} = 0$. This is illustrated in Fig. 8.3 (left).

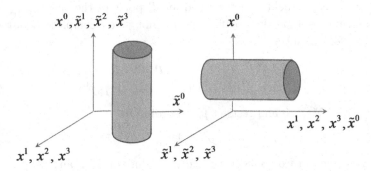

Fig. 8.3   Illustration of a bradyonic and tachyonic field localization. Under a superluminal transformation a bradyonic field localized within a space-like subspace $(x^1, x^2, x^3, \tilde{x}^0)$ (left) is transformed into a tachyonic field localized within a time like subscpece $(x^0, \tilde{x}^1, \tilde{x}^2, \tilde{x}^3)$ (right). The cases of a bradyon at rest and a tachyon with infinite speed are illustrated. For a nonvanishing bradyonic and a finite tachyonic speed the cylinders are suitably inclined.

Under a SLT the Klein-Gordon equation (8.35) becomes

$$(-\partial'^\mu \partial'_\mu + \tilde{\partial}'^\mu \tilde{\partial}'_\mu + M^2)\tilde{\phi}'(x'^\mu, \tilde{x}'^\mu) = 0. \tag{8.45}$$

This means that if in S a field satisfies the field equation (8.35), then the same field observed from S satisfies Eq. (8.45). The form of Eq. (8.45) is the same as the form of Eq. (8.35) in which $M^2 > 0$ is replaced with $M^2 < 0$, so that $-M^2 = \tilde{M}^2 > 0$, i.e.,

$$\left(\partial^\mu \partial_\mu - \tilde{\partial}^\mu \tilde{\partial}_\mu - \tilde{M}^2\right)\phi(x^\mu, \tilde{x}^\mu) = 0. \tag{8.46}$$

Its general solution is just like (8.37), but with x^μ, p^μ and $\tilde{x}^\mu$, $\tilde{p}^\mu$ interchanged:

$$\phi(x^\mu, \tilde{x}^\mu) = \int dp^0 d^3p\, d^3\tilde{p}\, \frac{1}{2\tilde{\omega}} \left[e^{i(-p_0 x^0 + \boldsymbol{px} + \tilde{\omega}\tilde{x}^0 - \tilde{\boldsymbol{p}}\tilde{\boldsymbol{x}})}\tilde{c}(p^0, \boldsymbol{p}, \tilde{\omega}, \tilde{\boldsymbol{p}}) \right.$$

$$\left. + e^{i(-p_0 x^0 + \boldsymbol{px} - \tilde{\omega}\tilde{x}^0 - \tilde{\boldsymbol{p}}\tilde{\boldsymbol{x}})}\tilde{c}(p^0, \boldsymbol{p}, -\tilde{\omega}, \tilde{\boldsymbol{p}}) \right], \tag{8.47}$$

where

$$\tilde{p}^0 = \pm\tilde{\omega} , \qquad \tilde{\omega} = |\sqrt{(p^0)^2 - \boldsymbol{p}^2 + \tilde{\boldsymbol{p}}^2 + \tilde{M}^2}|. \tag{8.48}$$

If we then choose

$$\tilde{c}(p^0, \boldsymbol{p}, \tilde{\omega}, \tilde{\boldsymbol{p}}) = \delta^3(\boldsymbol{p} - \boldsymbol{p}_c)\tilde{a}(p^0, \tilde{\omega}, \tilde{\boldsymbol{p}}),$$

$$\tilde{c}(p^0, \boldsymbol{p}, -\tilde{\omega}, \tilde{\boldsymbol{p}}) = \delta^3(\boldsymbol{p} - \boldsymbol{p}_c)\tilde{a}(p^0, -\tilde{\omega}, \tilde{\boldsymbol{p}}), \tag{8.49}$$

the field (8.47) becomes

$$\phi(x^\mu, \tilde{x}^\mu) = \int \mathrm{d}^3\tilde{p}\,\mathrm{d}p^0 \frac{1}{\tilde{\omega}_c} \left[e^{i\tilde{\omega}_c\tilde{x}^0} \tilde{a}(p^0, \tilde{\omega}, \tilde{\boldsymbol{p}}) \right.$$

$$\left. + e^{-i\tilde{\omega}_c\tilde{x}^0} \tilde{a}(p^0, -\tilde{\omega}, \tilde{\boldsymbol{p}}) \right] e^{i(-p_0 x^0 + \boldsymbol{p}_c\boldsymbol{x} - \tilde{\boldsymbol{p}}\tilde{\boldsymbol{x}})}. \tag{8.50}$$

The "initial" data can now be specified on the time-like 4-surface $\tilde{x}^0 = 0$, $\boldsymbol{x} = 0$, spanned by the coordinates $x^0, \tilde{x}^1, \tilde{x}^2, \tilde{x}^3$ (see Fig. 8.3 right). Within the space-like direction x^1, x^2, x^3 the field propagates as a plane wave $e^{i\boldsymbol{p}_c\boldsymbol{x}}$ and is thus not localized. It also is not localized within the space-like direction $\tilde{x}^0$, which occurs in the factor $e^{i\tilde{\omega}_c\tilde{x}^0}$.

It is straightforward to show [109] that for the particular choice

$$c(p^0, \boldsymbol{p}, \tilde{\omega}, \tilde{\boldsymbol{p}}) = \delta(p^2)\delta(p^3)\delta(\tilde{p}^0)\delta(\tilde{p}^1)a(p^0, \omega_x, \tilde{p}^2, \tilde{p}^3)$$

$$c(p^0, \boldsymbol{p}, -\tilde{\omega}, \tilde{\boldsymbol{p}}) = \delta(p^2)\delta(p^3)\delta(\tilde{p}^0)\delta(\tilde{p}^1)a(p^0, -\omega_x, \tilde{p}^2, \tilde{p}^3), \tag{8.51}$$

where $\omega_x = |\sqrt{(p^0)^2 + (\tilde{p}^0)^2 + (\tilde{p}^3)^2 + \tilde{M}^2}|$, Eq. (8.45) describes the tachyonic field considered in Sec. 8.2.

Within the framework of the space $M_{4,4}$ we thus have (citation from [109]): "a symmetry between bradyonic and tachyonic fields. None of those fields is more consistent than the other. Both kinds of fields are described by the ultrahyperbolic Klein-Gordon equation. We have reduced the problem of consistent propagating tachyonic fields to the problem of whether the ultrahyperbolic wave equation can make sense in physics. In addition, we have a problem of where do the extra dimensions $\tilde{x}^\mu$, $\mu = 0, 1, 2, 3$, come from and why do we not observe them. One possibility is just to suppose that our spacetime has not four, but eight dimensions (or, equivalently, that it is complex), and that the extra dimensions are not observed, because they are compactified. Another possibility is to consider the Clifford space, a $16D$ manifold whose tangent space at any of its points is the Clifford algebra of spacetime $M_{1,3}$."

8.4 Klein-Gordon equation in Clifford space

In Clifford space the Klein-Gordon equation reads [109]

$$\left(G^{MN}\partial_M\partial_N + M^2\right)\phi(x^M) = 0. \tag{8.52}$$

Using $x^M = (\sigma, x^\mu, x^{\mu\nu}, \tilde{x}^\mu, \tilde{\sigma})$, the explicit form of Eq. (8.52) is

$$\left(\frac{\partial^2}{\partial\sigma^2} + \partial^\mu\partial_\mu + \partial^{\mu\nu}\partial_{\mu\nu} - \tilde{\partial}^\mu\tilde{\partial}_\mu - \frac{\partial^2}{\partial\tilde{\sigma}^2} + M^2\right)\phi(\sigma, x^\mu, x^{\mu\nu}, \tilde{x}^\mu, \tilde{\sigma}) = 0. \tag{8.53}$$

The form of this equation is the same as of Eq. (8.35), except that the dimension of the space is now 16 and signature $(8,8)$. The coordinates x^μ and $\tilde{x}^\mu$ serve the same role as in Sec. 8.3. But in addition we now also have the extra eight coordinates, $x^{\mu\nu}$, σ and $\tilde{\sigma}$.

We will now exploit the fact that σ is time-like and $\tilde{\sigma}$ space-like, and combine them into

$$\tau = \frac{1}{\sqrt{2}}(\tilde{\sigma} - \sigma) , \qquad \lambda = \frac{1}{\sqrt{2}}(\tilde{\sigma} + \sigma) . \tag{8.54}$$

The new coordinates τ, λ are analogous to *light cone coordinates*.

Using (8.54) we can rewrite Eq. (8.52) as

$$\left(-2\frac{\partial^2}{\partial\tau\partial\lambda} + \partial^\mu\partial_\mu - \tilde{\partial}^\mu\tilde{\partial}_\mu + \partial^{\mu\nu}\partial_{\mu\nu} + M^2\right)\psi(\tau, \lambda, x^\mu, \tilde{x}^\mu, x^{\mu\nu}) = 0. \tag{8.55}$$

Denoting

$$x^{\bar{\mu}} = (x^\mu, \tilde{x}^\mu, x^{\mu\nu}) , \quad \partial_{\bar{\mu}} = (\partial_\mu, \tilde{\partial}_\mu, \partial_{\mu\nu}),, \quad \text{and} \quad \partial_\tau = \frac{\partial}{\partial\tau} , \quad \partial_\lambda = \frac{\partial}{\partial\lambda}, \tag{8.56}$$

Eq. (8.55) reads

$$(-2\partial_\tau\partial_\lambda + G^{\bar{\mu}\bar{\nu}}\partial_{\bar{\mu}}\partial_{\bar{\nu}} + M^2)\psi(\tau, \lambda, x^{\bar{\mu}}) = 0. \tag{8.57}$$

A general solution of the latter equation is

$$\psi = \int dp_\tau \, dp_\lambda d^{14}\bar{p} \; c(p_\tau, p_\lambda, p_{\bar{\mu}}) \, e^{i(p_\tau\tau + p_\lambda\lambda + p_{\bar{\mu}}x^{\bar{\mu}})} \delta(2p_\tau p_\lambda - G^{\bar{\mu}\bar{\nu}}p_{\bar{\mu}}p_{\bar{\nu}} + M^2). \tag{8.58}$$

Integrating the expression (8.58) over p_τ and introducing

$$\omega = p_\tau = \frac{1}{2p_\lambda}\left(G^{\bar{\mu}\bar{\nu}}p_{\bar{\mu}}p_{\bar{\nu}} - M^2\right), \tag{8.59}$$

$$a(\omega, p_\lambda, p_{\bar{\mu}}) \equiv \frac{1}{2p_\lambda} c(\omega, p_\lambda, p_{\bar{\mu}}) \tag{8.60}$$

we obtain

$$\psi(\tau, \lambda, x^{\bar\mu}) = \int \mathrm{d}p_\lambda \, \mathrm{d}^{14}\bar{p} \, a(\omega, p_\lambda, p_{\bar\mu}) e^{i\omega\tau} e^{ip_{\bar\mu} x^{\bar\mu}} e^{ip_\lambda \lambda}. \tag{8.61}$$

Because the relation (8.59) has no square root, all values of p_λ and $p_{\bar\mu}$ between $-\infty$ and ∞ give real ω, and thus take part in the integral (8.61). At initial $\tau = 0$ we have

$$\psi(\tau = 0, \lambda, x^{\bar\mu}) = \int \mathrm{d}p_\lambda \, \mathrm{d}^{14}\bar{p} \, a(\omega, p_\lambda, p_{\bar\mu}) e^{ip_{\bar\mu} x^{\bar\mu}} e^{ip_\lambda \lambda}. \tag{8.62}$$

From the latter equation one can obtain by the Fourier transformtaion the coefficients $a(\omega, p_\lambda, p_{\bar\mu})$ expressed in terms of $\psi(\tau = 0, \lambda, x^{\bar\mu})$. Hence from the initial data on a hypersurface $\tau = 0$ one can calculate the field at any τ by using the expression (8.62). The Cauchy problem in ultrahyperbolic spaces is thus well posed for a light-like hypersurface, which is known result [213].

In a particular setup in which the momentum p_λ has definite value Λ, so that

$$a(\omega, p_\lambda, p_{\bar\mu}) = \delta(p_\lambda - \Lambda) A(p_{\bar\mu}), \tag{8.63}$$

the wave function (8.61) takes the following form:

$$\psi(\tau, \lambda, x^{\bar\mu}) = \int \mathrm{d}^{14}\bar{p} \, A(p_{\bar\mu}) e^{i\omega\tau} e^{ip_{\bar\mu} x^{\bar\mu}} e^{i\Lambda\lambda} \equiv \varphi(\tau, x^{\bar\mu}) e^{i\Lambda\lambda}, \tag{8.64}$$

where

$$\omega = \frac{1}{2\Lambda} \left(G^{\bar\mu\bar\nu} p_{\bar\mu} p_{\bar\nu} - M^2 \right). \tag{8.65}$$

By plugging the Ansatz (8.65) into Eq. (8.55) we obtain the wave equation for $\varphi(\tau, x^{\bar\mu})$:

$$i \frac{\partial\varphi(\tau, x^{\bar\mu})}{\partial\tau} = \frac{1}{2\Lambda} (G^{\bar\mu\bar\nu} \partial_{\bar\mu} \partial_{\bar\nu} + M^2) \varphi(\tau, x^{\bar\mu}). \tag{8.66}$$

This is a generalization of the Stueckelbrg equation from 4D spacetime with coordinates x^μ, $\mu = 0, 1, 2, 3$ to the 14D space with coordinates $x^{\bar\mu} = x^\mu, \tilde{x}^\mu, x^{\mu\nu})$, where τ has the role of an evolution parameter. In our setup, based on Clifford space, the evolution parameter is given by a superposition (8.54) of the scalar and the pseudoscalar coordinate.

Equation (8.66) admits for a complex valued φ the conservation law

$$\frac{\partial\rho}{\partial t} + \partial_{\bar\mu} j^{\bar\mu} = 0, \tag{8.67}$$

where

$$\rho = \varphi^* \varphi \,, \quad j^{\bar{\mu}} = \frac{i}{2\Lambda} \left(\varphi^* \partial^{\bar{\mu}} \varphi - \partial^{\bar{\mu}} \varphi^* \varphi \right). \qquad (8.68)$$

The wave function is normalized according to $\int \mathrm{d}^{14} \bar{x} \, |\varphi(\tau, x^{\bar{\mu}})|^2 = 1$, and $|\varphi(\tau, x^{\bar{\mu}})|^2$ is the probability density of observing the "particle" at position $x^{\bar{\mu}}$.

The mass term M^2 in Eq. (8.66) gives a constant factor $\exp[-\frac{M^3 \tau}{2\Lambda}]$, so that a particular solutions is

$$\varphi_{(\bar{p})}(\tau, x^{\bar{\mu}}) = \mathrm{e}^{\frac{i}{2\Lambda} \left(G^{\bar{\mu}\bar{\nu}} p_{\bar{\mu}} p_{\bar{\nu}} + M^2 \right) \tau} \mathrm{e}^{i p_{\bar{\mu}} x^{\bar{\mu}}}. \qquad (8.69)$$

A general solution is an arbitrary superposition of the above terms, unconstrained in momenta $p_{\bar{\mu}}$. The phase factor with M^2 has no physical role, and it makes not much difference if $M^2 > 0$ or $M^2 < 0$. Solutions of Eq. (8.66) are thus not sensitive on whether the field satisfying the original equation (8.52) is bradyonic or tachyonic. In any case, the Stueckelberg equation (8.66) admits either subluminal or superluminal group velocities, defined from the dispersion relation (8.65):

$$v^{\bar{\mu}} = \frac{\mathrm{d}\omega}{\mathrm{d}p_{\bar{\mu}}} = \frac{p^{\bar{\mu}}}{\Lambda} \,, \quad p^{\bar{\mu}} = G^{\bar{\mu}\bar{\nu}} p_{\bar{\nu}}. \qquad (8.70)$$

Since $p^{\bar{\mu}}$ are unconstrained, it can be either $p^{\bar{\mu}} p_{\bar{\mu}} > 0$, or $p^{\bar{\mu}} p_{\bar{\mu}} < 0$, and hence the velocity (8.70) either subluminal or superluminal. In the presence of interactions that system can make a smooth transition from the subluminal to the superluminal velocity [112–114].

8.5 Tachyons and causality

If tachyons can indeed transmit information, as we have seen in Sec. 8.2, then they could be used for transmission of information into the past. This would presumably lead to the causal paradoxes. A person who received by means of a tachyonic antitelephone an information from the future about certain events could act so to cause a different sequence of events. For instance, if the tachyonic signals received today tell you that tomorrow you will be caught in a traffic jam, you could decide to take a different route and avoid the jam. Or suppose that you invented the tachyonic telephone and examine it in your laboratory. It is Monday, 9.00 a.m., and suddenly on the apparatus screen appears a message: "Helo, it is Monday, 12.05 p.m. I just told my boss that I succeeded in making a working "telephone" based on transmission of information by tachyons. He was not much impressed, on the contrary, he said that tachyones are impossible and that I abused my working time for non sense. Therefore I can no longer work in his lab and must find another job. Best, John."

This was of course a message that you wrote three hours later. So you decide not to tell your boss about your invention and thus keep your job. In the subsequent course of events you found clever ways to make your invention known to others. The situation just described is not paradoxical if the history is not fixed, but split. The two different courses of events are illustrated in Fig. 8.4.

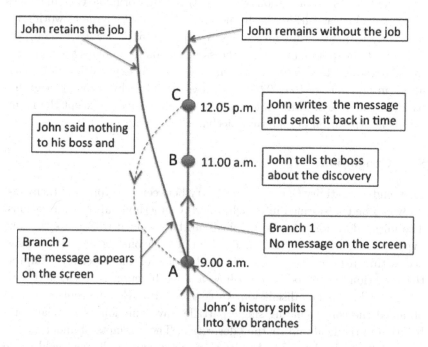

Fig. 8.4 If a tachyon signal is sent into the past, the history splits into two different histories.

Such scenario in which the history splits into two or more branches fits well into the Everett interpretation of quantum mechanics [215–221], also called many worlds interpretation. Namely, the wave function is a superposition of several or many possible outcomes of observation. All those outcomes exist in Hilbert space and are in this sense "real". For instance, in the case of the famous Schrödinger cat, the wave function is a superposition of live and dead cat. In the case of the situation with the tachyonic

telephone (Fig. 8.4) the wave function at point A is a superposition of the
state $|\Psi_1\rangle$ in which John observed the message from the future, and the
state $|\Psi_2\rangle$ in which he observed no message:

$$|\Psi_A\rangle = |\Psi_1\rangle + |\Psi_2\rangle. \tag{8.71}$$

The subsequent evolution of the branch $|\Psi_1\rangle$ is different than of the branch
$|\Psi_1\rangle$. Thus, if we take into account that the world is not deterministic
(as in classical physics), but quantum mechanical, then the scenarios with
information transmission into the past are not paradoxical.

An exhausting explanation why time travel by wormhole time machines
is not paradoxical within the framework of the Everett many worlds interpretation of quantum mechanics was provided by Deutsch [224]. In
Ref. [49] it was pointed out that the same reasoning resolves the so called
causal paradoxes associated with tachyons. An early explanation along such
lines can be found in Ref. [222]. The discovery of tachyons as propagating
particles—able to transmit information—would falsify all except the many
worlds interpetation of quantum mechanics.

8.6   Conclusion

In extended relativity there is a symmetry between bradyons and tachyons.
They can be transformed into each other by superluminal transformations.
The initial data for a bradyonic field can be specified within a space like
hypersurface, whilst the initial data for a tachyonic field can be specified within a time like hypersurface. In view of the preceding chapter,
the attention has to be paid to how a wave function is associated with
a given field. In our discussion we considered a Klein-Gordon field and
obtained the corresponding single particle wave function as a linear combination of the field and its time derivative. The attempts of understaning
tachyon (and also bradyon) localization in terms of a classical field alone
without considereing also its canonical momentum (time derivative) are
misconceived.

Bradyons moving in vacuum do not emit Čerenkov radiation. Therefore,
tachyons moving in vacuum also do *not* emit Čerenkov radiation. The
assertions in the literature that they do emit are incorrect from the point
of view of extended relativity. Consequently, superluminal neutrinos in
vacuum should not be expected to emit an analog of Čerenkov radiation
in the form of producing photons and e^+e^- pairs through Z^0 mediated

process, as claimed in Ref. [225]. Therefore, the absence of such a weak current analog of Čerenkov radiation in an experiment with neutrinos does not exclude the possibility of their superluminality. In this respect, the conclusion in Ref. [226] is not correct. Whether neutrinos within a certain beam move slower or faster than light has to be determined by measuring the time of flight.

Chapter 9

Ordering Ambiguity of Quantum Operators

A big stumbling block in our attempts to obtain consistent quantized versions of classical theories is the problem of ordering of operators. In simple cases, for instance in flat spaces, the order of quantum operators does not matter. However, if a space is curved, then the momentum operator acts on the position dependent metric, and there occurs the ambiguity concerning the order of operators. Such ambiguity typically occurs in the expressions for a Hamilton operator in curved space. In this chapter we discuss and further develop the finding of Ref. [227].

9.1 Introduction

An important ingredient of a theory is Hamiltonian. It usually contains a quadratic form composed of momenta and a metric:

$$H = H(p^2) , \qquad p^2 = p_a g^{ab} p_b , \qquad a, b = 1, 2, 3, ..., n. \tag{9.1}$$

For instance, in the case of a relativistic point particle, described by the reparametrization invariant action

$$I[X^\mu] = \int d\tau \left(\dot{X}^\mu \dot{X}_\mu \right)^{1/2} , \qquad \mu = 0, 1, 2, 3, \tag{9.2}$$

the Hamiltonian is

$$H(X^\mu, p^\mu) = p_\mu \dot{X}^\mu - L = p^\mu p_\mu - m^2 = 0, \tag{9.3}$$

which is a constraint among the canonical momenta containing $p^\mu p_\mu = p_\mu g^{\mu\nu} p_\nu$, i.e., a quadratic form that contains spacetime metric $g^{\mu\nu}$.

Alternatively, introducing a gauge $X^0 \equiv t = \tau$, $g_{0i} = 0$, the action is

$$I[X^i] = m \int dt \sqrt{g_{00} - q_{ij} \dot{X}^i \dot{X}^j}, \tag{9.4}$$

from which we obtain the Hamiltonian

$$H(X^i, p_i) = \sqrt{g_{00}} \sqrt{m^2 + q^{ij} p_i p_j},$$

(9.5)

where $q_{ij} = -g_{ij}$, $q^{ik} q_{jk} = \delta^i{}_j$. In general, the metric depends on position in the configuration space over which the Hamiltonian is defined. In classical theory the expression (9.1) is unambiguous. On the contrary, in the quantized theory momenta are operators satisfying

$$[\hat{x}^a, \hat{p}_b] = i\delta^a{}_b, \quad [\hat{x}^a, \hat{x}^b] = 0, \quad [\hat{p}^a, \hat{p}^b] = 0.$$

(9.6)

In coordinate representation, in which $\hat{x}^a$ is diagonal, the operators are represented according to

$$\hat{x}^a \longrightarrow \langle x|\hat{x}^a|x'\rangle = x^a \delta^n(x - x'),$$

(9.7)

$$\hat{p}_a \to \langle x|\hat{p}^a|x'\rangle = -i\partial_a \delta(x, x') + F_a(x)\delta(x, x'),$$

(9.8)

where

$$\delta(x, x') = \frac{\delta(x - x')}{\sqrt{g(x)}} = \frac{\delta(x - x')}{\sqrt{g(x')}}$$

(9.9)

is the invariant δ function. If the determinant $\det g_{ab}$ of the metric is negative, then $g(x)$ denotes the determinant multiplied by minus one, i.e., $g(x) = -\det g_{ab} > 0$.

The basis states are normalized according to

$$\langle x|x'\rangle = \delta(x, x'),$$

(9.10)

which is an invariant relation, valid in any coordinates.

The requirement that the matrix elements representing $\hat{p}_a$ be Hermitian, namely

$$\langle x|\hat{p}^a|x'\rangle^* = \langle x'|\hat{p}^a|x\rangle,$$

(9.11)

together with $[\hat{x}^a, \hat{p}_b] = 0$, restricts the choice of $F_a(x)$ to

$$F_a(x) = \partial_a \left(-\frac{1}{4} \ln g - \chi \right),$$

(9.12)

where $\chi(x)$ is an arbitrary function. For the choice $\chi = 0$, it is

$$\langle x|\hat{p}^a|x'\rangle = -i \left(\partial_a + \frac{1}{4} \partial_a \ln g \right) \delta(x, x'),$$

(9.13)

where

$$\frac{1}{4}\partial_a \ln g = \frac{1}{4} g^{-1}\partial_a g = \frac{1}{2} g^{-1/2}\partial_a g^{1/2} = g^{-1/4}\partial_a g^{1/4} = \frac{1}{2}\Gamma^b_{ab}.$$

(9.14)

The definition of the quantum Hamilton operator, corresponding to the classical Hamiltonian (9.1), is thus ambiguous. In the literature are used various ordering prescriptions.

9.2 Geometric definition of momentum

We are now going to show how ordering ambiguities do not arise if we define momentum as a vector

$$p = \gamma^a p_a. \tag{9.15}$$

Here $\gamma^a(x)$ are basis vectors, satisfying[1]

$$\gamma^a \cdot \gamma^b \equiv \frac{1}{2}\left(\gamma^a\gamma^b + \gamma^b\gamma^a\right) = g^{ab}(x). \tag{9.16}$$

The Hamiltonian contains just the square of the momentum, namely

$$p^2 = (\gamma^a p_a)^2 = (\gamma^a p_a)(\gamma^b p_b). \tag{9.17}$$

In the classical theory γ^a and p_a commute, therefore p^2 is equal to the quadratic form (2.1).

When passing to the quantized theory, the momentum (9.15) becomes the momentum vector operator

$$\hat{p} = \gamma^a \hat{p}_a = -i\gamma^a\partial_a \equiv -i\partial. \tag{9.18}$$

The quantum Hamiltonian then contains

$$\hat{p}^2 = (\gamma^a\hat{p}_a)^2 = (-i\gamma^a\partial_a)(-i\gamma^b\partial_b) = -\partial^2. \tag{9.19}$$

There is no ambiguity of how to form the square of the vector momentum operator. In the following we will calculate $\hat{p}^2$ explicitly.

The *vector derivative* or *gradient* is defined as

$$\partial = \gamma^a\partial_a, \tag{9.20}$$

where ∂_a is an operator, called derivative, that can act on a scalar, vector, bivector, or any other Clifford algebra valued field.

Let in general be

$$\Phi = \varphi(x)\mathbf{1} + \varphi^b(x)\gamma_b + \varphi^{b_1 b_2}(x)\gamma_{b_1} \wedge \gamma_{b_2} + \ldots + \varphi^{b_1 b_2 \ldots b_n}(x)\gamma_{b_1} \wedge \gamma_{b_2} \wedge \ldots \wedge \gamma_{b_n}. \tag{9.21}$$

Then

$$\partial_a\Phi = \partial_a\varphi(x)\mathbf{1} + \partial_a\varphi^b(x)\gamma_b + \varphi^b(x)\partial_a\gamma_b$$

$$+ \partial_a\varphi^{b_1 b_2}(x)\gamma_{b_1} \wedge \gamma_{b_2} + \varphi^{b_1 b_2}(x)\partial_a(\gamma_{b_1} \wedge \gamma_{b_2})$$

$$+ \partial_a\varphi^{b_1 b_2 \ldots b_n}(x)\gamma_{b_1} \wedge \gamma_{b_2} \wedge \ldots \wedge \gamma_{b_n} + \varphi^{b_1 b_2 \ldots b_n}(x)\partial_a(\gamma_{b_1} \wedge \gamma_{b_2} \wedge \ldots \wedge \gamma_{b_n}). \tag{9.22}$$

---

[1] Here $\gamma^a(x) = e^a{}_{\bar{a}}(x)\gamma^{\bar{a}}$, where $\gamma^{\bar{a}} \cdot \gamma^{\bar{b}} = \delta^{\bar{a}\bar{b}}$ if the tangent space at the point x is euclidean, otherwise instead of $\delta^{\bar{a}\bar{b}}$ it stays a pseudo Euclidean metric, e.g., $\eta^{\bar{a}\bar{b}} = \text{diag}(1, 1, \ldots, -1, -1, \ldots)$.

In the above expression the wedge products of gammas form the basis of the Clifford algebra, whereas the field $\varphi(x)$, φ^b, $\varphi^{b_1 b_2}(x)$, ..., are scalar component[2].

If acting on a scalar field, such as $\varphi(x)$ or φ^b, the derivative ∂_a behaves as the partial derivative:

$$\partial_a \varphi = \frac{\partial \varphi}{\partial x^a}, \qquad \partial_a \varphi^b = \frac{\partial \varphi^b}{\partial x^a}. \tag{9.23}$$

Acting on a vector γ_b, it gives

$$\partial_a \gamma_b = \Gamma_{ab}^c \gamma_c, \tag{9.24}$$

where Γ_{ab}^c is the connection. Similarly, for a reciprocal vector $\gamma^b = g^{bc}\gamma_c$, we have

$$\partial_a \gamma^b = -\Gamma_{ac}^b \gamma^c. \tag{9.25}$$

When acting on a vector field $\varphi^b \gamma_b$, which is a superposition of the component fields φ^b and the basis tangent vectors γ_a, the derivative gives

$$\partial_a (\varphi^b \gamma_b) = \partial_a \varphi^b \gamma_b + \varphi^b \partial_a \gamma_b = \left(\partial_a \varphi^b + \Gamma_{ac}^b \varphi^c \right) \gamma_b = D_a \varphi^b \gamma_b, \tag{9.26}$$

where $D_a \varphi^b = \partial_a \varphi^b + \Gamma_{ac}^b \varphi^c$ is *the covariant derivative* of the tensor calculus.

The vector derivative ∂ acting on a vector field thus gives

$$\partial(\varphi^b \gamma_b) = \gamma^a \partial_a (\varphi^b \gamma_b) = \gamma^a \gamma_b D_a \varphi^b = \gamma^a \gamma^b D_a \varphi_b = \left(\gamma^a \cdot \gamma^b + \gamma^a \wedge \gamma^b \right) D_a \varphi_b. \tag{9.27}$$

Here $\gamma^a \cdot \gamma^b$ is the symmetric part (9.16) and

$$\gamma^a \wedge \gamma^b = \frac{1}{2} \left(\gamma^a \gamma^b - \gamma^a \gamma^b \right) \tag{9.28}$$

the antisymmetric part of *the Clifford product*

$$\gamma^a \gamma^b = \gamma^a \cdot \gamma^b + \gamma^a \wedge \gamma^b. \tag{9.29}$$

For the product of two vector derivatives acting on a vector we have

$$\partial \partial (\varphi^c \gamma_c) = (\gamma^a \partial_a)(\gamma^b i \partial_b)(\varphi^c \gamma_c) = \gamma^a \gamma^b \gamma^c D_a D_b \varphi_c, \tag{9.30}$$

and in general,

$$\partial^m (\varphi^b \gamma_b) = \gamma^{a_1} \gamma^{a_2} ... \gamma^{a_m} \gamma^b D_{a_1} D_{a_2} ... D_{a_m} \varphi_b. \tag{9.31}$$

In such way we can determine the action of the vector derivative or its powers on any Clifford algebra valued field occurring in the expansion (9.21).

---

[2]In tensor calculus, $\varphi(x)$, φ^b, $\varphi^{b_1 b_2}(x)$, ... are called "scalar", "vector", and, in general, tensor fields of rank $r = 0, 1, 2, ..., n$. But within Clifford algebra they are all scalars.

Of special interest for our purpose is to see what happens if the vector derivative acts twice on a *scalar field*. We find

$$\partial\partial\varphi = (\gamma^a\partial_a)(\gamma^b\partial_b)\varphi = \gamma^a\gamma^b D_a D_b\varphi. \tag{9.32}$$

Notice that $D_a\varphi = \partial_a\varphi$, so that $(\partial_a\partial_b - \partial_b\partial_a)\varphi = 0$. Then $(D_a D_b - D_b D_a)\varphi = -(\Gamma^c_{ab} - \Gamma^c_{ba})\varphi = 0$, which vanishes if the connection is symmetric. In a torsionless manifold the torsion $C^c_{ab} = \Gamma^c_{ab} - \Gamma^c_{ba}$ vanishes, and we have

$$\partial\partial\varphi = g^{ab} D_a D_b\varphi = D_a D^a\varphi = \frac{1}{\sqrt{g}}\partial_a\left(\sqrt{g}\,g^{ab}\partial_b\varphi\right). \tag{9.33}$$

In the case of a vector field, Eq. (9.30) gives

$$\partial\partial(\varphi^c\gamma_c) = (\gamma^a \cdot \gamma^b + \gamma^a \wedge \gamma^b) D_a D_b\varphi^c\gamma_c. \tag{9.34}$$

Multiplying the latter equation from the left by γ^d and taking the scalar part, we obtain

$$\gamma^d\partial\partial(\varphi^a\gamma_a) = D_a D^a\varphi^d. \tag{9.35}$$

Besides the scalar and vector field, an important role in quantum field theory have *spinor fields*. As shown in Chapter 4, a spinor is an element of a left or right ideal of Clifford algebra[3]:

$$\Psi = \psi^\alpha\xi_\alpha. \tag{9.36}$$

Here ξ_α are basis spinors, defined in Chapter 4 and ψ^α spinor components fields. The action of the derivative on ξ_α gives

$$\partial_a\xi_\alpha = \Gamma^\beta_{a\alpha}\xi_\beta, \tag{9.37}$$

where the superposition coefficients $\Gamma^\beta_{a\alpha}$ form the *spin connection*. For the reciprocal spinor basis, $\xi^\alpha = z^{\alpha\beta}\xi_\beta$, where $z^{\alpha\beta}$ is the spinor metric, we have

$$\partial_a\xi^\alpha = -\Gamma^\alpha_{a\beta}\xi^\beta. \tag{9.38}$$

Using (9.7) and (9.37), we obtain

$$\partial_a\Psi = \left(\partial_a\psi^\alpha + \Gamma^\beta_{a\alpha}\psi^\beta\right)\xi_\alpha = D_a\psi^\alpha\xi_\alpha. \tag{9.39}$$

The Dirac equation is thus

$$i\gamma^a\partial_a\Psi + m\Psi = (i\gamma^a D_a + m)\,\psi^\alpha\xi_\alpha = 0. \tag{9.40}$$

If we multiply Eq. (9.40) from the left by ψ^β and take the scalar part, we obtain the component form of the Dirac equation

$$i\left((\gamma^a)^\beta_{\;\alpha} + m\delta^\beta_{\;\alpha}\right)\psi^\alpha, \tag{9.41}$$

---

[3]For more details see also Refs. [60, 68].

where

$$(\gamma^a)^{\beta}{}_{\alpha} = \langle \xi^{\beta} \gamma^a \xi_{\alpha} \rangle_S \qquad (9.42)$$

are the Dirac matrices.

Equipped with the connections (9.24), (9.25), (9.37) and (9.38) we can calculate the double action of the vector derivative on a spinor field as well. We find

$$\partial \partial \Psi = (\gamma^a \partial_a)(\gamma^b \partial_b) \Psi = g^{ab} D_a D_b \psi^{\alpha} \xi_{\alpha} + \gamma^a \wedge \gamma^b R_{ab}{}^{\alpha}{}_{\beta} \psi^{\beta} \xi_{\alpha}, \qquad (9.43)$$

where

$$g^{ab} D_a D_b \psi^{\alpha} = g^{ab} \left(\partial_a D_b \psi^{\alpha} - \Gamma^c_{ab} D_c \psi^{\alpha} + \Gamma^{\alpha}_{a\beta} D_b \psi^{\beta} \right), \qquad (9.44)$$

and

$$R_{ab}{}^{\alpha}{}_{\beta} = \partial_a \Gamma^{\alpha}_{b\beta} - \partial_b \Gamma^{\alpha}_{a\beta} + \Gamma^{\delta}_{a\beta} \Gamma^{\alpha}_{b\delta} - \Gamma^{\delta}_{b\beta} \Gamma^{\alpha}_{a\delta} \qquad (9.45)$$

is the curvature tensor expressed in term of the spin connections. In Eq. (9.44) we see how the covariant derivative acts on the vector indices a, b and the spinor indices α, β.

Multiplying Eq. (9.43) from the left by ξ^{δ} and taking the scalar part, we obtain the following equation:

$$\langle \xi^{\delta} \partial \partial \Psi \rangle_S = g^{ab} D_a D_b \psi^{\delta} + \langle \xi^{\delta} \gamma^a \wedge \gamma^b \xi_{\alpha} \rangle_S R_{ab}{}^{\alpha}{}_{\beta} \psi^{\beta} \qquad (9.46)$$

where

$$\langle \xi^{\delta} \gamma^a \wedge \gamma^b \xi_{\alpha} \rangle_S = \frac{1}{2} [\gamma^a, \gamma^b]^{\delta}{}_{\alpha} \equiv (\sigma^{ab})^{\delta}{}_{\alpha} \qquad (9.47)$$

are matrix elements of the of the commutator of gammas which is proportional to the spin tensor.

We see that because the square of the vector momentum $\partial^2 = \partial \partial = (\gamma^a \partial_a)^2$ is unambiguously defined, there is no ambiguity of how to form in curved space the Klein-Gordon equation for a scalar and a vector field. There is also no ambiguity concerning the Dirac equation and how to square it: in addition to the Klein-Gordon part, one obtains a coupling term between the curvature tensor and the spin tensor σ^{ab}, which results also in the usual procedure. But in the geometric treatment presented here, there are no other curvature dependent terms, like R, which occur in the usual, non geometric approaches that rely on various possible choices of ordering prescription.

Matrix elements of the vector momentum operator in the x-representation are

$$\langle x | \hat{p} | x' \rangle = -i \gamma^a(x) \partial_a \delta(x, x'). \qquad (9.48)$$

Using

$$\partial_a \delta(x, x') = \partial'_a \delta(x, x') - \delta(x, x') \frac{1}{g(x')} \partial'_a \sqrt{g(x')} \qquad (9.49)$$

and

$$\Gamma^b_{ab} = \frac{1}{g} \partial'_a \sqrt{g}, \qquad (9.50)$$

we find

$$-i\gamma^a(x)\partial_\delta(x, x') = i\gamma^a(x')\partial'_a \delta(x, x'). \qquad (9.51)$$

Therefore it follows that

$$\langle x'|\hat{p}|x\rangle^* = \langle x|\hat{p}|x'\rangle, \qquad (9.52)$$

which is the Hermiticity condition for the matrix elements of the operator ∂.

In curved space, the completeness relation is[4]

$$\int |x\rangle \sqrt{g(x)} \mathrm{d}x \langle x| = 1. \qquad (9.53)$$

Consequently, the action of $\hat{p}$ on a state vector $|\Phi\rangle$ gives

$$\langle x|\hat{p}|\Phi\rangle = \int \langle x|\hat{p}|x'\rangle \sqrt{g(x')} \mathrm{d}x' \langle x'|\Phi\rangle = -i\gamma^a \partial_a \Phi = \hat{p}\Phi, \qquad (9.54)$$

where $\langle x|\Phi\rangle = \Phi(x)$ is a wave function, which can be a scalar, vector, or any higher grade component of the Clifford algebra valued field (9.21), or it can be a spinor field (9.36).

The double action of the vector momentum operator on a state $|\Phi\rangle$ gives

$$\langle x|\hat{p}^2|\Phi\rangle = \int \langle x|\hat{p}|x'\rangle \sqrt{g(x')} \mathrm{d}x' \langle x'|\hat{p}|x''\rangle \sqrt{g(x'')} \mathrm{d}x'' \langle x''|\Phi\rangle. \qquad (9.55)$$

To calculate this expression, we observe that from Eqs. (9.25) and (9.50) it follows that

$$\partial_a \left(\gamma^a \sqrt{g} \right) = 0. \qquad (9.56)$$

Using the latter relation, performing partial integration, we arrive at the result

$$\langle x|\hat{p}^2|\Phi\rangle = -\gamma^a \partial_a (\gamma^b \partial_b \Phi) = -D_a D^a \Phi. \qquad (9.57)$$

This demonstrates that the product of two vector differential operators $\hat{p} = -i\gamma^a \partial_a$, acting on a wave function, is equal to the matrix element $\langle x|\hat{p}^2|\Phi\rangle$ which can be calculated by inserting twice the complete set of the position eigenstates satisfying the completeness relation (9.53).

---

[4]Namely,
$\langle x|x'\rangle = \int \langle x|x''\rangle \sqrt{g(x'')} \mathrm{d}x'' \langle x''|x'\rangle = \int \frac{1}{g(x)} \delta^n(x - x'') \sqrt{g(x'')} \mathrm{d}x'' \frac{1}{g(x'')} \delta^n(x'' - x')$
$= \frac{1}{g(x)} \delta^n(x - x') = \delta(x, x').$

9.3　The equations of motion for the expectation value of momentum

Let us focus our attention to the case of the scalar component φ of Φ (see Eq. (9.21)), assume that it is complex valued, depends on an evolution parameter τ and the coordinates x^a, $a = 1, 2, 3, ..., n$, is normalized according to

$$\int \mathrm{d}^n x \, \sqrt{g} \varphi^*(\tau, x) \varphi(\tau, x) = 1, \tag{9.58}$$

and satisfies the Schrödinger equation

$$i \frac{\partial \varphi}{\partial \tau} = H \varphi. \tag{9.59}$$

The Hamilton operator is a function of p^2, such as that of Eq. (9.3) or (9.5). The expectation value of the operator $\hat{p} = -i \gamma^a \partial_a$ is

$$\langle \hat{p} \rangle = -i \int \mathrm{d}^n x \, \sqrt{g} \varphi^* \gamma^a \partial_a \varphi. \tag{9.60}$$

Taking its complex conjugate, we have

$$\langle \hat{p} \rangle^* = i \int \mathrm{d}^n x \sqrt{g} \, \varphi \gamma^a \partial_a \varphi^*$$

$$= -i \int \mathrm{d}^n x \sqrt{g} \, \varphi^* \gamma^a \partial_a \varphi$$

$$- \int \mathrm{d}^n x \varphi^* \partial_a \left(\sqrt{g} \gamma^a \right) \varphi + i \int \mathrm{d}^n x \, \partial_a \left(\varphi^* \sqrt{g} \gamma^a \varphi \right). \tag{9.61}$$

Omitting the surface term and using Eq. (9.56), we obtain that the expectation value is real:

$$\langle \hat{p} \rangle^* = \langle \hat{p} \rangle. \tag{9.62}$$

This means that the vector momentum operator is self adjoint with respect to the scalar product (9.58).

The derivative of the expectation value (9.60) with respect to τ is

$$\frac{\mathrm{d} \langle \hat{p} \rangle}{\mathrm{d} \tau} = \int \mathrm{d}^n x \sqrt{g} \, \frac{\partial \varphi^*}{\partial \tau} \hat{p} \varphi + \varphi^* \hat{p} \frac{\partial \varphi}{\partial \tau}$$

$$= i \int \mathrm{d}^n x \sqrt{g} \, \left((H \varphi^*) \hat{p} \varphi - \varphi^* \hat{p} H \varphi \right). \tag{9.63}$$

By taking into account that $H(\hat{p}^2)$ is a series of powers of $\hat{p}^2$ and using Eq. (9.57) we obtain

$$\int \mathrm{d}^n x \, \sqrt{g} (H \varphi^*) \hat{p} \varphi = \int \mathrm{d}^n x \sqrt{g} \, \varphi^* H \hat{p} \varphi. \tag{9.64}$$

Therefore Eq. (9.63) becomes

$$\frac{d\langle \hat{p} \rangle}{d\tau} = i \int d^n x \sqrt{g}\, \varphi^* [H, \hat{p}] \varphi. \tag{9.65}$$

If $H = H(\hat{p}^2)$, then $[H, \hat{p}] = 0$, and we have

$$\frac{d\langle \hat{p} \rangle}{d\tau} = 0. \tag{9.66}$$

This means that the expectation value of the vector momentum operator does not change with τ.

To find what precisely is the meaning of Eq. (9.66), let us examine how the expectation value $\langle \hat{p} \rangle$ is calculated. From Eq. (9.60) we see that it is calculated as the integral of the vector momentum operator sandwiched between φ^* and φ, that is, as the integral of a vector $p(x') = \varphi^*(x')(-i\gamma^a(x')\partial'_a\varphi(x')$ over the manifold. The result of such integral is a vector at a chosen point x (see Ref. [227]). Namely, all vectors $p(x')$ at different points x' of the manifold are brought together to the same point x, where they are summed. The vectors are brought together by the parallel transport along the geodesic joining x' and x. The result of the integration is thus a vectors

$$P = \gamma^a(x)P_a(x). \tag{9.67}$$

A question arises as to which point x shall we choice. Let us choose for x the expectation value $\langle \hat{x} \rangle \equiv X(\tau)$ of the particle's position, which in fact is different at every τ, therefore also the expected momentum P is taken at different points $X(\tau)$. The comparison of P at different points can be done by means of the geometric derivative ∂_a. As we have seen, ∂_a performs the parallel transport of a vector from the point $x^a + \delta x^a$ to the point x^a. Therefore,

$$P(x + \delta x) - P(x) = \delta P = \partial_a P \delta x^a. \tag{9.68}$$

Taking $x^a = X^a(\tau)$ and $x^a + \delta x^a = X^a(\tau + \delta\tau)$, we have $\delta x^a = \dot{X}^a \delta\tau$. Consequently, Eq. (9.68) reads

$$\delta P = \partial_a P \dot{X}^a \delta\tau, \tag{9.69}$$

or

$$\frac{dP}{d\tau} = \partial_a P \dot{X}^a. \tag{9.70}$$

If we now take into account Eqs. (9.24), (9.66) and (9.67) we obtain

$$\left(\partial_a P^c + \Gamma^c_{ad} P^d \right) \dot{X}^a = 0. \tag{9.71}$$

Here $\dot{X}^a$ are components of the velocity vector $\gamma_a \dot{X}^a$ which is the expectation value of the operator $\frac{1}{2}(\gamma_a \dot{\hat{x}}^a + \dot{\hat{x}}^a \gamma_a)$:

$$\gamma_a \dot{X}^a = \langle \frac{1}{2}(\gamma_a \dot{\hat{x}}^a + \dot{\hat{x}}^a \gamma_a) \rangle. \tag{9.72}$$

From the Heisenberg equations

$$\dot{\hat{x}} = i[H, \hat{x}^a] = \frac{i}{2}[\hat{p}^2, \hat{x}^a]$$

$$= -i \left(g^{ab} \partial_b + \frac{1}{2} \gamma^b \partial_b \gamma^a \right)$$

$$= -i \left(\partial^a - \frac{1}{2} \gamma^b \Gamma^a_{bc} \gamma^c \right), \tag{9.73}$$

we have

$$\gamma_a \dot{\hat{x}}^a = -i \left(\gamma^a \partial_a - \frac{1}{2} g^{bc} \Gamma^a_{bc} \gamma_a \right), \tag{9.74}$$

and

$$\dot{\hat{x}}^a \gamma_a = -i \left(\gamma^a \partial_a + g^{ab} \partial_b \gamma_a + \frac{1}{2} \gamma^b \partial_b \gamma^a \gamma_a \right)$$

$$= -i \left(\gamma^a \partial_a + \frac{1}{2} g^{bc} \Gamma^a_{bc} \gamma_a \right). \tag{9.75}$$

Therefore,

$$\frac{1}{2}(\gamma_a \dot{\hat{x}}^a + \dot{\hat{x}}^a \gamma_a) = -i \gamma^a \partial_a = \hat{p}, \tag{9.76}$$

and

$$\gamma_a \dot{X}^a = P = \gamma_a P^a. \tag{9.77}$$

The symmetrized velocity operator is thus equal to the vector momentum operator $\hat{p}$, and its expectation value $\gamma_a \dot{X}^a$ is equal to the expected momentum P.

Inserting Eq. (9.77) into Eq. (9.71), gives

$$\partial_a \dot{X}^c \dot{X}^a + \Gamma^c_{ad} \dot{X}^d \dot{X}^a = \frac{d\dot{X}^c}{d\tau} + \Gamma^c_{ad} \dot{X}^d \dot{X}^a = 0, \tag{9.78}$$

which is the equation of geodesic. The expected momentum $P = \langle \gamma^a \hat{p}_a \rangle = \gamma_a \dot{X}^a$ thus follows a geodesic trajectory in curved space, i.e., at every τ it is tangent to a geodesic.

9.4 On the integration of vectors in curved space

A common wisdom is that vectors cannot be integrated in curved space. This is true for *components* of vectors, but not for vectors as *geometric objects*. Geometrically, a vector a is an object that in a given basis consists of components a^a and basis vectors γ_a, $a = 1, 2, ..., n$. Vectors can be elegantly represented as grade 1 elements of a Clifford algebra $Cl(n)$, generated by γ_a, satisfying (9.16). In general, γ_a are not orthonormal, and the metric $g_{ab}(x)$ depends on position x. Locally, at every position x one can introduce an orthonormal basis $\gamma_{\bar{a}}$, $\bar{a} = 1, 2, ..., n$, satisfying

$$\gamma_{\bar{a}} \cdot \gamma_{\bar{b}} = \frac{1}{2}(\gamma_{\bar{a}}\gamma_{\bar{b}} + \gamma_{\bar{b}}\gamma_b a) = g_{\bar{a}\bar{b}}, \tag{9.79}$$

where the metric $g_{\bar{a}\bar{b}} = \text{diag}(1, 1, 1, ..., -1, -1, ...)$ is diagonal with as signature (p, q), $p + q = n$. The basis vectors $\gamma_a(x)$ we will call *world basis vectors*, and $\gamma_{\bar{a}}$ *local orthonormal vectors*. At every point x we have the relation

$$\gamma_a = (\gamma_a \cdot \gamma^{\bar{a}})\gamma_{\bar{a}} \equiv e_a{}^{\bar{a}}\gamma_{\bar{a}}, \tag{9.80}$$

and the inverse relation

$$\gamma_{\bar{a}} = (\gamma_{\bar{a}} \cdot \gamma^a)\gamma_a \equiv e_{\bar{a}}{}^a\gamma_a. \tag{9.81}$$

Here $e_a{}^{\bar{a}}$ is the vielbein, and $e_{\bar{a}}{}^a$ its inverse, satisfying

$$e_a{}^{\bar{a}}e_{\bar{a}}{}^b = \delta_a{}^b, \quad e_a{}^{\bar{a}}e_{\bar{a}b} = g_{ab}. \tag{9.82}$$

An arbitrary vector at a point x is then

$$a(x) = a^a(x)\gamma_a(x) = a^a(x)e_a{}^{\bar{b}}(x)\gamma_{\bar{b}}(x) \equiv a^{\bar{b}}(x)\gamma_{\bar{b}}(x). \tag{9.83}$$

A vector at a nearby point x_1 with coordinates $x_1^a + \epsilon^a$ is then

$$a(x_1) = a^a(x_1)\gamma_a(x_1) = \left(a^a(x) + \partial_b a^a(x)\epsilon^b + ...\right)\left(\gamma_a(x) + \partial_b\gamma_a(x)\epsilon^b + ...\right)$$

$$= a^a(x)\gamma_a(x) + \partial_b a^a(x)\epsilon^b\gamma_a(x) + \partial_b\gamma_a(x)\epsilon^b + ...$$

$$= \left(a^a(x) + \partial_b a^a(x)\epsilon^b + \Gamma^a_{bc}(x)a^c(x)\epsilon^b + ...\right)\gamma_a(x)$$

$$= \left(a^a(x) + D_b a^a(x)\epsilon^b + ...\right)\gamma_a(x) = a^a(x_1, x)\gamma_a(x), \tag{9.84}$$

where $D_b a^c \equiv \partial_b a^a + \Gamma^c_{bc}a^c$ is the covariant derivative. The quantities in the bracket are the components $a^a(x_1, x)$ of the vector, obtained by the parallel transport of the vector $a(x_1)$ from the point x_1 to the point x.

Using

$$a^a(x_1) = a^a(x) + \partial_b a^a(x)\epsilon^b, \qquad (9.85)$$

we find that

$$a^a(x_1, x) = a^b(x_1)g_b{}^a(x_1, x), \qquad (9.86)$$

where

$$g_b{}^a(x_1, x) = \delta_b{}^a + \Gamma^a_{bc}\epsilon^c , \qquad \epsilon^c = x_1^c - x^c. \qquad (9.87)$$

The difference between $a(x_1)$ and $a(x)$ is thus given by the covariant derivative

$$a(x_1) - a(x) = D_b a^a(x)\gamma_a(x)\epsilon^b. \qquad (9.88)$$

The sum of $a(x)$ and $a(x_1)$ is

$$a(x) + a(x_1) = \left(2a(x) + D_b a^a(x)\epsilon^b\right)\gamma_a(x). \qquad (9.89)$$

We see that if we take into account not only the components a^a, but also the basis vectors γ_a, then we can make sums and differences of vectors at different points of a manifold. If the points are not infinitesimally close to each other, then a path between the points matters. This is so, because we can repeat the procedure

$$a(x_1) = a(x) + \partial_b a(x)\epsilon^b(x) \qquad (9.90)$$

for the nearby points x_1 and x_2, and obtain

$$\begin{aligned}
a(x_2) &= a(x_1) + \partial a(x_1)\epsilon^b(x_1) \\
&= a(x) + \partial_b a(x)\epsilon^b(x) + \partial_b\left(a(x) + \partial_c a(x)\epsilon^c(x)\right)\epsilon^b(x_1) \\
&= \left[a^a(x) + D_b a^a(x)\left(\epsilon^b(x) + \epsilon^b(x_1)\right) + D_b D_c a^a(x)\epsilon^b(x)\epsilon^c(x_1)\right]\gamma_a(x).
\end{aligned} \qquad (9.91)$$

The same we can do for the points x_2, x_3, etc. At each step, the infinitesimal vector ϵ^b is different. So we proceed along a curve that joins two points of the manifold, say, x and x', at which we compare the vectors.

The finite relation can be written formally as

$$a(x') = a^a(x')\gamma_a(x') = a^b(x')g^a{}_b(x', x)\gamma_a(x) = a^a(x', x)\gamma_a(x), \qquad (9.92)$$

where $g^a{}_b(x', x)$ is the *parallel propagator* from x' to x along a given curve. The quantities $a^a(x', x)$ are the vector components, obtained by the parallel transport of the vector $a(x')$ from the point x' to the point x *along a specified curve which, from now on, will be taken to be the geodesic.*

Let us now consider the integral of a vector field $a(x)$ over a domain Ω within the manifold:

$$\int d^n x \sqrt{g(x)} a^a(x) \gamma_a(x). \tag{9.93}$$

First, we observe that such integral is invariant under general coordinate transformations. Second, *choosing a fixed point* x', and using the relation (9.92), in which we now rename x into x', and vice versa, we have

$$\int d^n x \sqrt{g(x)} a^a(x) \gamma_a(x) = \int d^n x \sqrt{g(x)} g^a{}_b(x, x') a^b(x) \gamma_a(x')$$

$$= A^a(x') \gamma_a(x') = A(x'). \tag{9.94}$$

The result of such integration is a vector $A(x') = A^a(x') \gamma_a(x')$ at a chosen point x'. The components $A^a(x') = \int d^n x \sqrt{g(x)} g^a{}_b(x, x') a^b(x)$ are obtained by the summation (the integration over x) of the components $a^a(x, x') = g^a{}_b(x, x') a^b(x)$ of the vectors $a(x, x')$, obtained by the parallel transported of the vectors $a(x)$, $x \in \Omega$, from x to x'. In other words, by construction the integral (9.94) is such that all vectors are transported along the geodesics from x to a chosen point x', where they are integrated.

Using (9.92) and (9.81), we have

$$\gamma_{\bar{a}} = e_{\bar{a}}{}^a(x) \gamma_a(x) = e_{\bar{a}}{}^b(x) g_b{}^a(x, x') \gamma_a(x') = e_{\bar{a}}{}^b(x, x') \gamma_a(x'), \tag{9.95}$$

where

$$e_{\bar{a}}{}^b(x, x') = e_{\bar{a}}{}^b(x) g_b{}^a(x, x') \tag{9.96}$$

is the vielbein transported from x to x'.

Inverting (9.96), we obtain the following expression for the parallel propagator:

$$g_b{}^a(x, x') = e_b{}^{\bar{a}}(x) e_{\bar{a}}{}^a(x, x'). \tag{9.97}$$

Equation (9.92), now written with the x and x' interchanged,

$$a^a(x) \gamma_a(x) = a^b(x) g_b{}^a(x, x') \gamma_a(x') \tag{9.98}$$

can be interpreted in two different ways:

(i) Components $a^a(x)$ are transported from x to x' to become

$$a^a(x, x') = a^b(x) g_b{}^a(x, x'). \tag{9.99}$$

This interpretation we adopted so far.

(ii) Basis vectors $\gamma_a(x')$ are transported from x' to x to become

$$\gamma_b(x, x') = g_b{}^a(x, x') \gamma_a(x') = \gamma_a(x). \tag{9.100}$$

According to this second interpretation the basis vectors $\gamma_a(x)$ occurring in the left hand side of Eq. (9.98) are in fact the transported basis vectors.

Multiplying Eq. (9.100) by $e_{\bar{b}}{}^{b}(x)$, and using $\gamma_{\bar{b}}(x,x') = e_{\bar{b}}{}^{b}(x)\gamma_b(x,x')$, we obtain the relation for the parallel transport of $\gamma_{\bar{b}}$:

$$\gamma_{\bar{b}}(x,x') = g_{\bar{b}}{}^{\bar{a}}(x,x')\gamma_{\bar{a}}(x'), \qquad (9.101)$$

where

$$g_{\bar{b}}{}^{\bar{a}}(x,x') = e_{\bar{b}}{}^{b}(x)g_b{}^{a}(x,x')e_a{}^{\bar{a}}(x'). \qquad (9.102)$$

By defining vectors in geometric term as elements of a Clifford algebra, the integration of vectors in curved spaces thus makes perfect sense. The result of such integration is a vector at a chosen point of the manifold.

9.5   Conclusion

We have again seen how useful are the approaches based on Clifford algebras. The momentum operator $\hat{p}_a = -i\partial_a$ is not a geometric object. To obtain a geometric object one has to involve basis vectors γ^a and replace $\hat{p}_a$ with $\hat{p} = -i\gamma^a\partial_a$, which is the vector momentum operator. We represented basis vectors by the generators of Clifford algebra and extended the definition of the operator ∂_a so to act both on the scalar components of a Clifford number and also to its Clifford algebra basis elements. In particular, acting with the vector momentum operator twice on a scalar field, there is no ambiguity about the order of operators, even in curved spaces. Namely, in a curved space the basis vectors determine the metric according to $\gamma_a \cdot \gamma_b = g_{ab}$. The action of the derivative ∂_a on γ^a gives the connection according to Eq. (9.24). The double action of the vector momentum operator on a scalar field thus gives $g^{ab}D_aD_b\varphi$ as shown in Eq. (9.33). A similar expression we obtain for a vector field (see Eq. (9.35)). In both cases no curvature scalar takes place in the expressions. The situation is different, if $\hat{p}$ acts on a spinor field. Then we obtain Eq. (9.43) which contains a term with the curvature tensor.

We have also considered the equations of motion for the expectation value of the vector momentum and found that they imply the geodesic equation (9.71). For this purpose it was necessary to consider integrals over vector fields in curved spaces. Such integration is generally considered as ill defined. But again, by employing Clifford numbers, we were able to give sense to such integration and thus to the expectation value of the vector momentum and its components.

The resolution of ordering ambiguities by employing vector momentum operators can be extended from finite dimensional to infinite dimensional spaces. In such a manner, the famous Wheeler-DeWitt equation could be formulated as a mathematically well defined expression. Also in this respect canonical quantum gravity would then become a viable physical theory, after we have already shown that its higher derivative extension, necessary to render the theory renormalizable, though containing negative energies, is not plugged with instabilities.

Chapter 10

What Have We Learned?

We started with the social behavior in chimpanzee's communities in which, as experiments show, inventions are mostly achieved by the individuals of low-rank status and thus usually do not spread within the community. Such studies aim at a better understanding of social behavior in humans by studying social interactions in the less-involved chimpanzee's communities. We see that as far as inventions or new discoveries are concerned, people are often not much better than monkeys, and sometimes, people are worse. Typically, a new out-of-the-box discovery is ignored, ridiculed or strongly opposed by the rest of the community. In the mildest case, such a discovery is ignored, and this happens if the discoverer is young and unknown; otherwise, the person usually faces ridicule and violent opposition.

In a scientific community, social behavior has additional components, which are nicely discussed in the book *The Trouble with Physics* by Lee Smolin [1]. Among others, Smolin analyzes indepth why, in the last at that time two or three (now four) decades, theoretical physics has not produced new results of fundamental significance. The current academic system does not encourage young scientists with their own research projects. A young theoretical physicist must join an existing major project to have a chance for a successful scientific career. That most young theoretical physicists join an existing research project is not bad in itself. A problem arises if nobody or very few individuals with their own ideas can start working in an academic institution. Smolin observes that there are two types of scientists, namely, seers and crafts people, and that science needs both: "Seers are good at asking genuinely novel but relevant questions, [...], and have the ability to look at the state of a technical field and see a hidden assumption or a new avenue of research". Unfortunately, academic institutions do not embrace such creative rebels with this rare talent. Mostly, academic institutions exclude these individuals. Smolin observes that currently, there is a place only for "craftspeople" who usually master their discipline better than seers but have no genuinely new ideas of their own. A consequence is that fundamental theoretical physics is no longer advancing as it had been during the last two centuries when every twenty years or so, there was a major breakthrough discovery.

In the book, Smolin lucidly discusses the sociology that has led to such a state of affairs. Let me add here my own opinion, or perhaps an observation, that the progress was slowed when people started to consider too seriously the number of citations, received shortly after publication, as a measure of the importance of an article. To have a chance at rapidly attracting many citations, a scientific paper must focus on a subject or a field that is being investigated by many researchers, that is, a paper has to be about hot topics. The journals today favor the publication of papers that are of "general scientific interest", which in fact means that their subject must fit into one of the major fields that are being investigated at the time of submission The policy is clear: such a paper is likely to quickly attract citations and thus contribute to the journal's "impact factor". Today, the journals, according to their policy, exclude papers written by "seers", whose pioneering investigation is by definition not yet of general scientific interest because seers investigate topics that have not yet been investigated at all. The journals that compete for a high impact factor are thus no longer serving the development of fundamental theoretical physics (in particular) and to science (in general) as well as they could. A paper of general scientific interest can only be an incremental advancement to what has already been investigated by many others. This practice is not bad in itself, bad is that a "revolutionary" paper cannot be published at all in this current climate. Therefore, as I perceive it, the invention of impact factor, and especially its usage as a measure of the importance of a journal, has led to such a "sociology" within the community of theoretical physicists that progress has stalled.

To point out that impact factor cannot be taken too seriously as a measure of a journal's importance for the development of science, I am now asking: "What was the impact factor of the not-so-well-known journal of Brno (German: Brunn) in which Mendel published his laws of genetics"? His paper was appreciated only thirty-five years later by a renown scientist, who made Mendel's work known to the scientific community. The impact factor (that takes into account citations within the two-year period only) of that relatively obscure journal was certainly substantially lower than the "impact factor" of the renowned journals in Mendel's time. However, given the fundamental importance and influence of Mendel's paper for genetics, it would be absurd to say that the impact of that journal to science was as insignificant as its impact factor would suggest. Mendel published his paper in a relatively unknown journal, obtained very few citations in the first thirty-five years, and yet, as it turned out, this was one of the greatest papers of all times.

The impact factor has yet another damaging consequence. Many people read only the journals with high impact factors and ignore those with low impact factors. However, today, an important revolutionary paper is very likely to be rejected by renowned journals and eventually appear in a journal with a low impact factor. But since those journals have not many readers, such a paper will remain unnoticed for a long time[1]. Fortunately, we have also preprint servers, such as arXiv. But scientists outside academic institutions[2] cannot post their work there unless they obtain an endorsement from an established scientist. So nowadays a "new Einstein" has a really difficult time to be noticed at all. Of course, most would-be new Einsteins are cranks, but some of them, especially those with a university education in the field, might be right. Therefore, a good system should have a mechanism for identifying such persons, like a good detector must have the ability to detect rare particles and not just classify them as noise.

Hence, a big stumbling block against unification—of interactions on the one hand, and of gravity and quantum theory, on the other hand— is the way how physics research is organized. Therefore, a new insight reaching far out of the box cannot attract attention. Moreover, numerous persisting fixed conceptions cemented within physics community present impenetrable obstacles against the progress. In this book, I describe some of those stumbling blocks.

One of the greatest obstacles is in considering negative energies that occur in spaces with indefinite signature (ultrahyperbolic spaces) and that are, in particular, associated with higher derivative theories as being something bad, leading to instabilities in the systems described by such theories. If a system possesses degrees of freedom with positive and negative energies, then in the presence of coupling terms, they can feed each other so that initially small amplitudes start growing. With the interaction potential that people usually have in mind, namely, unbounded either from above or

---

[1] Notice the parallel with the observation that inventions, since they are usually achieved by low-ranking individuals, do not spread within a chimpanzee community.

[2] Recall that seers mostly cannot work in academic institutions.

below (or both), the amplitudes grow towards infinity. Such growth cannot happen in the presence of a bounded potential. I have shown this in many examples of calculations performed with Mathematica and found that the system was always stable. This is a crucial observation that opens the door to ultrahyperbolic spaces and higher derivative theories, including higher derivative gravity, as viable ingredients of physics. I have then pointed out how to deal with the notorious instantaneous vacuum decay occurring in quantum field theories in the presence of negative energy states. I have shown not only that such a vacuum decay is not bad, but also that it has important physical consequences that need to be further explored in detail. One of them is that it provides a possible mechanism for the Big Bang. The other consequence is the formation of the Dirac sea of negative energy states of fermions.

This concept is connected to the fact that spinors can be considered as elements of the left (or right) ideals of a Clifford algebra. In the case of the Clifford algebra $Cl(1,3)$ of the four-dimensional spacetime, there are four left ideals. The usual particles that we observe belong to the first left ideal, while the remaining three ideals are associated with unobserved particles that can assume the role of dark and mirror matter. The $Cl(1,3)$ has signature $(8,8)$ and is thus an ultrahyperbolic space. The field theory built over such space possesses negative energies, and its vacuum is thus prompted to instantaneous decay. This setup can lead to a particular sort of incomplete vacuum decay such that in the first left ideal, there is a sea of negative energy fermions, while in the third left ideal, there is the sea of positive energy fermions, with the remaining states being unfilled. In this scenario, the sector of $Cl(1,3)$ associated with the particles that we observe is the first left minimal ideal, in which we thus have positive energy fermions and the Dirac sea of negative energy fermions.

By considering spinors and vectors as elements of a Clifford algebra, the usual view that those distinct objects transform differently, turns out to be not quite correct. Under a given transformation within, say, $Cl(1,3)$, vectors and spinors transform according to the same rule. It is only the manifestation of such rule that is different, but not the rule itself. For instance if we act on a vector by an appropriate operator R from the left and by R^{-1} from the right, the result is another vector, which is, in general, a superposition of the original vectors. If we do the same with a spinor, e.g., of a first left minimal ideal of $Cl(1,3)$, the result is not an element of the same ideal, but an object that, in general, is a superposition of the spinors of all four ideals. However, we can act on a spinor with R from the left only, in which case we obtain a spinor of the same ideal. The usual wording that vectors transform according to $v \to v' = RvR^{-1}$, and spinor according to $\psi \to \psi' = R\psi$, refers only to the transformations that bring vectors into vectors and spinors into spinors. Yet, this does not imply that on a vector one cannot act according to Rv, and on a spinor, according to $R\psi R^{-1}$. The object Rv is no longer a vector, but a superposition of scalars, vectors, bivectors, pseudovectors and pseudoscalar, while $R\psi R^{-1}$ is no longer a spinor of a given ideal, but any Clifford number.

Of special interest are discrete transformations, including space inversion. We know how it acts on vectors and what a mirror picture looks like. For consistency, we have to act with the same transformation on spinors. After such a transformation, a spinor of the first ideal transforms into a spinor of a third ideal. Mirror particles thus do not live in the same ideal as we do, but in another ideal, and interact with mirror gauge fields; therefore, they are invisible to us. In this way, the mirror symmetry is an exact symmetry of nature. The possibility of restoring mirror symmetry by introducing mirror particles was considered by Lee and Yang in the same paper in which they questioned parity conservation. With Clifford algebra and spinors as Clifford numbers, we have a theoretical description of mirror particles and an exact parity symmetry. This description sheds new light on the plausible viability of the theory of everything proposed by G. Lisi [229], which was rejected by Distler [228] just because it implied "antigeneration" of mirror fermions and because it was not clear how to explain chiral behavior in the processes with weak interactions.

An important topic in modern physics is strings and branes. Although string theory has run into problems, it remains important for the progress of physics, in spite of the fact that, so far, it has had no direct contact with observations. String and brane theories, as currently formulated, need revision and extension to incorporate additional concepts. We have seen that an important concept is brane space, i.e., the configuration space of branes. A big problem that theorists have encountered so far in attempting to formulate brane theory in an infinite-dimensional space was the reparameterization invariance. The same brane can be described in an infinite number of ways, therefore, it is difficult to take this fact into account when envisaging branes as points in a brane space. However, as shown in Ref. [49] such a problem can be circumvented by introducing the concept of a kinematical brane space $\mathcal{M}$, whose objects are all branes, including those related by a reparameterization. By introducing a suitable metric in such a brane space, one obtains, among many other possibilities, the familiar

Dirac-Nambu-Goto branes. According to such a view, the familiar string and brane theory is just a special case of a general brane theory, with a particular choice of metric in brane space. A different choice of metric gives a different brane theory, and since there are infinitely many possible brane space metrics, there are infinitely many possible variants of the familiar string and brane theory. However, it was also pointed out [49] that not all kinematically possible metrics of $\mathcal{M}$ are also dynamically possible metrics. A dynamically possible metric of $\mathcal{M}$ must satisfy certain dynamical equations of motion, e.g., Einstein's equations in $\mathcal{M}$.

Another possible avenue to brane theory is described in this book. We start from the concept of a *flat brane* and *flat brane space*. A flat brane can be envisaged and described as a continuous collection of interaction-free point particles, while a flat brane space is a particular brane space whose metric is a generalization of Euclidean or pseudo Euclidean metric to infinite dimensions. Hence, as Minkowski space is a particular kind of spacetime, such that its metric can be cast into the diagonal form $\eta_{\mu\nu}$, so flat brane space is a particular kind of brane space $\mathcal{M}$, such that its metric can be cast into the form $\eta_{\mu\nu}\delta^{p+1}(\xi, \xi')$, $\mu, \nu = 0, 1, 2, ..., D$. A flat brane, as a collection of particles, can be described upon quantization as a set of fields, e.g., scalar fields. I have shown that by introducing particular local interactions among the fields, the expectation value of the momentum operator satisfies the classical equations of motion for the Dirac-Nambu-Goto brane, including the Nambu-Goto string. To show this, we also need to employ the concept of a position operator and to calculate its expectation value with respect to a wave packet state of a collection of point particles.

We can thus consider a quantized brane as a cluster of interacting quantum fields. To show that such a system indeed corresponds to the classical Dirac-Nambu-Goto brane, we need the concept of a position operator. In the literature, such a concept is considered problematic, but I have shown that within the relativistic quantum field theory, the concept of a single-particle wave packet in position space makes sense (despite that within the framework of relativistic quantum mechanics alone, without recourse to relativistic quantum field theory, it can indeed be problematic if negative energies are incorrectly taken into account and if the meaning of the wave function is improperly ascribed to the field satisfying the Klein-Gordon equation). An entire chapter is devoted to the position operator, to its covariant formulation, and to properly understanding the meaning and behavior of wave packet states in position space. The confusion regarding particle position in relativistic quantum theories and quantum field theories is one of the greatest stumbling blocks of unification. Once this persistent block is removed, a way becomes opened to quantized branes, and consequently, to quantum gravity, if we adopt the braneworld scenario according to which our universe is a brane.

Perhaps the greatest block on our way to unification stands on the seeming restrictions imposed by special relativity, a theory that is by now more than hundred years old. According to the usual interpretation, the theory of relativity forbids faster-than-light motion. In 1970s and 1980s, there was substantial interest in including tachyons into the framework of relativity or into an extended version of it. At that time there was also much confusion and misunderstanding, but the subject crystallized in the works of E. Recami and his collaborators [198, 199], who formulated *extended relativity* (see also [197]). In four-dimensional spacetime the theory needs complex coordinates, while in its simplified version in 2D spacetime, the coordinates are real. This relation suggests that in a six dimensional spacetime with equal number of time-like and space-like dimensions, i.e., with signature $(3, 3)$, one can extend relativity by using only real coordinates [208–211].

In the current book, I discuss those approaches to extended relativity and connect them to the concept of Clifford space, which is also a manifold with a neutral signature, namely, $(8, 8)$. I have shown that in a manifold with a neutral signature, we can have propagating superluminal wave packets. Thus, the usual arguments saying that tachyonic fields cannot propagate with superluminal speeds do not hold within the framework of extended relativity in 6D spacetime or 16D Clifford space. However, once the issue related to the possibility of superluminal wave packets is resolved, there still remains the problem of causality violation. After all, propagating tachyons would enable sending signals into the past and hence "change" the past. At this point, the limitations of a classical theory reveal their full weakness. The world is not classical and deterministic. The world is quantum and indeterministic. Although it is true that "Nobody really understands quantum mechanics" [214], among the many interpretations of quantum mechanics, there exists one that resolves the so-called causality problems connected with signaling, and traveling [49, 222–224] into the past.

The discovery of tachyons would be a crucial experiment that would boost the advance of theoretical physics. Above all, it would require the adoption of an Everett-like interpretation of quantum mechanics [49, 215–217]. Next, it would be necessary to place tachyons into an appropriate theoretical framework, which very likely is an extended relativity in 6D spacetime or in 16D Clifford space. Thus, after a hundred years, we would have finally come beyond the theory of relativity. This theory would not be rejected but only extended. In September 2011, the OPERA collaboration [230] announced the detection of faster-than-light neutrinos produced at CERN, Geneva, and recorded in Gran Sasso, Italy. A burst of theoretical activity followed, proposing various explanations for the occurrence of superluminal neutrinos.

All this enthusiasm about finally advancing physics at the fundamental level was shut down when OPERA revealed [231, 232] that they had found a loosely plugged cable in the equipment and that, with the cable properly fixed, the observed neutrinos were not superluminal. A great relief was felt by the majority of the physics community, who were happy that the theory of relativity was "saved" and again proven to be correct. Given that during last forty years, nothing truly new and exciting has happened in fundamental physics, one would expect disappointment within the physics community, not satisfaction. It is no wonder that fundamental (theoretical) physics no longer advances if most physicists are happy if the findings of a would-be breakthrough experiment, performed by a respectable group, turns out to be false.

The OPERA incident will very likely block the advance of physics even more because any other group that in the future might detect evidence of superluminal propagation, would probably discard their experimental results as theoretically impossible and say nothing about them. After all, not all scientists are Shechtmans[3], willing to expose themselves to ridicule and exclusion. Shall we remain happy with ascribing the theory of relativity the status of an immutable religion and be, at the same time, unhappy that fundamental physics no longer advances?

A message of this book is that if we really wish genuine progress in fundamental physics, then we should become more open-minded, willing to step out of the box, and seriously contemplate when and whether the circumstances require us to do so. We should be cautious when facing a fresh, unconventional view on a subject about which we already have a firm conviction because it is a textbook "truth". Although most such ideas are wrong or "even not wrong" and obvious nonsense, some of the "crazy" ideas that contradict textbook wisdom are correct. We should pay special attention to the unconventional proposals of professional physicists and not

---

[3]See Chapter 1.

plainly reject them without (at least) a second thought. I have identified some crucial topics, which have so far been considered completely understood and being unquestionable foundations that have to be taken into account when constructing new theories. I have shown why the prevailing firm convictions about negative energies, ultrahyperbolic spaces, spinor transformations, relativistic particle localization, tachyons, causality, and ordering ambiguities of quantum operators are stumbling block against unification. Clarification of those persistent misconceptions in physics is necessary for its further advancement at fundamental level.

Bibliography

[1] L. Smolin, *The Trouble with Physics* (Penguin Books, London, 2008).

[2] R. Kendal, A. Whiten, S. F. Brosnan, S. P. Lambeth, S. J. Schapiro and W. Hoppitt, *Evol. Hum. Behav.* **36** (1), 6572 (2015). doi: 10.1016/j.evolhumbehav.2014.09.002

[3] V. Horner, D. Proctor, K. E. Bonnie, A. Whiten and F. B. M. de Waal, *PLoS One* **5** (5), e10625 (2010). Published online 2010 May 19. doi: 10.1371/journal.pone.0010625

[4] Galileo Galilei, *Dialogo sopra i Massimi sistemi del Mondo*, published in 1632; For English translation see, e.g. *Dialogues Concerning Two New Sciences*, Translated from Italian and Latin into English by Henry Crew and Alfonso de Salvio. With an Introduction by Antonio Favaro (University of Adelaide, 2014; This translation first published by Macmillan, 1914).

[5] R. P. Feynman, R. B Leighton and M. Sands, *Feynman Lectures on Physics Vol. 2* (Addison-Wesley Publishing Company, 1963).

[6] F. Rohrlich, "The Saga of 3/4", in *The Physicist's Conception of Nature*, edited by J. Mehra (Reidel, Dordrecht, 1973), pp. 339–344.

[7] E. Fermi, *Physikalische Zeitschrift* **23**, 340 (1922).

[8] E. Fermi, *Rev. Mod. Phys.* **4**, 87 (1932).

[9] M. Pavšič, *Adv. Appl. Clifford Algebras* **22**, 449 (2012) [arXiv:1104.2266 [math-ph]].

[10] M. V. Ostrogradski, *Mem. Acad. Imper. Sci. St. Petersbg.* **6**, 385 (1850).

[11] M. Pavšič, *J. Phys. Conf. Ser.* **437**, 012006 (2013) [arXiv:1210.6820 [hep-th]].

[12] M. Pavšič, *Int. J. Geom. Meth. Mod. Phys.* **13**, no. 09, 1630015 (2016) [arXiv:1607.06589 [gr-qc]].

[13] A. Pais and G.E. Uhlenbeck, *Phys. Rev.* **79**, 145 (1950).

[14] A. Mostafazadeh, *Phys. Lett. A* **375**, 93 (2010) [arXiv:1008.4678 [hep-th]].

[15] R. Banerjee, *New (Ghost-Free) Formulation of the Pais-Uhlenbeck Oscillator*, arXiv:1308.4854 [hep-th].

[16] K. Bolonek, P. Kosiński, *Acta Phys. Polon.* **36**, 2115 (2005).

[17] E.V. Damaskinsky and M.A. Sokolov, *J. Phys. A: Math. Gen.* **39**, 10499 (2006).

[18] K. Bolonek and P. Kosinski, *J. Phys. A* **40**, 11561 (2007) [arXiv:quant-ph/0612091].

[19] F. Bagarello, *Int. J. Theor. Phys.* **50**, 3241 (2011).

[20] M. C. Nucci and P. G. L. Leach, *Phys. Scripta* **81**, 055003 (2010) [arXiv:0810.5772 [math-ph]].

[21] I. Masterov, *Nucl. Phys. B* **902**, 95 (2016) [arXiv:1505.02583 [hep-th]].

[22] I. Masterov, *Nucl. Phys. B* **907**, 495 (2016) [arXiv:1603.07727 [math-ph]].

[23] A. Déctor, H. A. Morales-Técotl, L. F. Urrutia and J. D. Vergara, *Coping with the Pais-Uhlenbeck oscillator's ghosts in a canonical approach*, arXiv:0807.1520 [quant-ph].

[24] S. Pramanik and S. Ghosh, *Mod. Phys. Lett. A* **28**, 1350038 (2013) [arXiv:1205.3333 [math-ph]].

[25] A. V. Smilga, *Phys. Lett. B* **632**, 433 (2006) [arXiv:hep-th/0503213].

[26] A. V. Smilga, *Nucl. Phys. B* **706**, 598 (2005) [hep-th/0407231].

[27] A. V. Smilga, *SIGMA* **5**, 017 (2009) [arXiv:0808.0139 [quant-ph]].

[28] M. Pavšič, *Mod. Phys. Lett. A* **28**, 1350165 (2013) [arXiv:1302.5257 [gr-qc]].

[29] D. Robert and A.V. Smilga, *J. Math. Phys.* **49**, 042104 (2008).

[30] H. Bondi, *Rev. Mod. Phys.* **29**, 423 (1957).

[31] Shanshan Yao, Xiaoming Zhou and Gengkai Hu, *New J. Phys.* **10**, 043020 (2008).

[32] P. Sheng, X. X. Zhang, Z. Y. Liu and C. T. Chan, *Physica B* **338**, 201 (2003).

[33] J. Mei, Z. Y. Liu, W. J. Wen and P. Sheng, *Phys. Rev. Lett.* **96**, 024301 (2006).

[34] J. Mei, Z. Y. Liu, W. J. Wen and P. Sheng, *Phys. Rev. B* **76**, 134205 (2007).

[35] H. Choi and P. Rudra, *Pair Creation Model of the Universe From Positive and Negative Energy*, (2014) [http://vixra.org/abs/1403.0180].

[36] H. Choi, *Hypothesis of Dark Matter and Dark Energy with Negative Mass*, (2009), [http://vixra.org/abs/0907.0015].

[37] H. Choi, *On Problems and Solutions of General Relativity* (Commemoration of the 100th Anniversary of General Relativity)", (2015), [http://vixra.org/abs/1511.0240].

[38] M. Wimmer, A. Regensburger, C. Bersch, Mohammad-Ali Miri, S. Batz, G. Onishchukov, D. N. Christodoulides and U. Peschel, *Nature Physics* **9**, 780 (2013)

[39] D. S. Kaparulin, S. L. Lyakhovich and A. A. Sharapov, *Eur. Phys. J. C* **74**, 3072 (2014) [arXiv:1407.8481 [hep-th]].

[40] D. S. Kaparulin, S. L. Lyakhovich and A. A. Sharapov, *J. Phys. A* **49**, 155204 (2016) [arXiv:1510.08365 [hep-th]].

[41] D. S. Kaparulin, I. Y. Karataeva and S. L. Lyakhovich, *Eur. Phys. J. C* **75**, 552 (2015) [arXiv:1510.02007 [hep-th]].

[42] D. S. Kaparulin and S. L. Lyakhovich, *Russ. Phys. J.* **57**, no. 11, 1561 (2015).

[43] D. Cangemi, R. Jackiw and B. Zwiebach, *Annals of Physics* **245**, 408 (1996).

[44] E. Benedict, R. Jackiw and H. J. Lee, *Phys. Rev. D* **54**, 6213 (1996).

[45] M. Pavšič, *Phys. Lett. A* **254**, 119 (1999) [arXiv:hep-th/9812123].

[46] R. P. Woodard, *Lect. Notes Phys.* **720**, 403 (2007) [arXiv:astro-ph/0601672].

[47] M. Pavšič *Found. Phys.* **35**, 1617 (2005) [arXiv:hep-th/0501222].

[48] M. Pavšič, *Quantum Vacuum and Clifford Algebras*, Invited talk presented at Cosmology and the Quantum Vacuum 2015 19-25 June 2015 Rhodes, Greece

[49] M. Pavšič, *The Landscape of Theoretical Physics: A Global View; From Point Particles to the Brane World and Beyond, in Search of a Unifying Principle* (Kluwer, 2001) [arXiv:gr-qc/0610061].

[50] M. Tegmark, *Fortschritte der Physik* **46**, 855 (1998).

[51] H. Dieter Zeh, *Z. Naturforsch A* **71**, 195 (2016).

[52] E. C. G. Sudarshan and B. Misra, *J. Math. Phys.* **18**, 763 (1977).

[53] Z. Albert, Y. Aharonov and S. D'Amato, *Phys. Rev. Lett.* **54**, 5 (1985).

[54] Y. Aharonov, S. Massar, S. Popescu and L. Vaidman, *Phys. Rev. Lett.* **77**, 983 (1996).

[55] S. M. Carroll, M. Hoffman and M. Trodden, *Phys. Rev. D* **68**, 023509 (2003) [arXiv:astro-ph/0301273].

[56] E. Cartan, *Leçons sur la théorie des spineurs I & II* (Paris: Hermann, 1938); E. Cartan *The theory of spinors*, English transl. by R.F. Streater (Paris: Hermann, 1966); C. Chevalley *The algebraic theory of spinors* (New York: Columbia U.P, 1954); I. M. Benn, R. W. Tucker, *An introduction to spinors and geometry with appliccations in physics* (Bristol: Hilger, 1987).

[57] P. Budinich, *Phys. Rep.* **137**, 35 (1986); P. Budinich and A. Trautman, *Lett. Math. Phys.* **11**, 315 (1986).

[58] S. Giler, P. Kosiński, J. Rembieliński and P. Maślanka, *Acta Phys. Pol. B* **18**, 713 (1987).

[59] J. O. Winnberg, *J. Math. Phys.* **18**, 625 (1977).

[60] M. Pavšič, *Phys. Lett. B* **692**, 212 (2010) [arXiv:1005.1500 [hep-th]].

[61] M. Budinich, *J. Math. Phys.* **50**, 053514 (2009); M. Budinich, *J. Phys. A* **47**, 115201 (2014); M. Budinich, *Adv. Appl. Clifford Algebras* **25**, 771 (2015).

[62] E. Kähler, *Rendiconti di Matematica* **21**, 425 (1962).

[63] S. I. Kruglov, *Int. J. Theor. Phys.* **41**, 653 (2002).

[64] D. Spehled and G. C. Marques, *Eur. Phys. J.* **61**, 75 (2009).

[65] M. Pavšič, "Beyond Spacetime: On the Clifford Algebra Based Generalization of Relativity", *Proceedings of the 5th Mathematical Physics Meeting*, July 6–17, 2008, Belgrade, Serbia (Ed. B. Dragović, Z. Rakić, Institute of Physics, Belgrade 2009), p. 343

[66] M. Pavšič, *J. Phys. A* **41**, 332001 (2008) [arXiv:0806.4365 [hep-th]].

[67] M. Pavšič, "Geometric Spinors, Generalized Dirac Equation and Mirror Particles," in *Theoretical Physics and its new Applications* (Selected contributions at the 3rd International Conference on Theoretical Physics, 24-28 Jun 2013. Moscow, Russia) (Moscow, 2014), p. 61. arXiv:1310.6566 [hep-th],

[68] M. Pavšič, *Adv. Appl. Clifford Algebras* **22**, 449 (2012) [arXiv:1104.2266 [math-ph]].

[69] F. Piazzese, in *Clifford Algebras and their Applications to Mathematical Physics*, F. Bracks et al. (eds.) (Kluwer Academic Publishers, 1993), p. 325.

[70] M. Pavšič, *Phys. Lett. B* **614**, 85 (2005) doi:10.1016/j.physletb.2005.03.052 [hep-th/0412255].

[71] M. Pavšič, *Int. J. Mod. Phys. A* **21**, 5905 (2006) [arXiv:gr-qc/0507053].

[72] T. D. Lee and C. N. Yang, *Phys. Rev.* **104**, 254 (1956).

[73] I. Yu. Kobzarev, L. B. Okun and I. Ya. Pomeranchuk, *Soviet J. Nucl. Phys.* **5**, 837 (1966).

[74] M. Pavšič, *Int. J. Theor. Phys.* **9**, 229 (1974).

[75] E. W. Kolb, D. Seckel, M. S. Turner, *Nature* **314**, 415 (1985).

[76] S. I. Blinnikov and M. Y. Khlopov, *Sov. J. Nucl. Phys.* **36**, 472 (1982) [*Yad. Fiz.* **36**, 809 (1982)].

[77] S. I. and M. Y. Khlopov, *Sov. Astron.* **27**, 371 (1983) [*Astron. Zh.* **60**, 632 (1983)].

[78] R. Foot, H. Lew and R. R. Volkas *Phys. Lett. B* **272**, 67 (1991).

[79] R. Foot, H. Lew and R. R. Volkas *Mod. Phys. Lett. A* **7**, 2567 (1992).

[80] R. Foot, *Mod. Phys. Lett.* **9**, 169 (1994).

[81] R. Foot and R. R. Volkas, *Phys. Rev. D* **52**, 6595 (1995).

[82] H. M. Hodges, *Phys. Rev. D* **47**, 456 (1993).

[83] R. Foot, *Phys. Lett. B* **452**, 83 (1999).

[84] R. Foot, *Phys. Lett. B* **471**, 191 (1999).

[85] R.N. Mohapatra, *Phys. Rev. D* **62**, 063506 (2000).

[86] Z. Berezhiani, D. Comelli and F. Villante, *Phys. Lett. B* **503**, 362 (2001).

[87] P. Ciarcelluti, *Int. J. Mod. Phys. D* **14**, 187 (2005).

[88] P. Ciarcelluti, *Int. J. Mod. Phys. D* **14**, 223 (2005).

[89] P. Ciarcelluti and R. Foot, *Phys. Lett. B* **679**, 278 (2009).

[90] R. Foot, *Int. J. Mod. Phys. A* **29**, 1430013 (2014) [arXiv:1401.3965 [astro-ph.O]].

[91] M. Pavšič, "Quantized Fields à la Clifford and Unification", in *Beyond Peacefull Coexistence; The Emergence of Space, Time and Quantum*", Ed. I. Licata (Imperial College Press, 2016).

[92] P. A. M. Dirac, *The Principles of Quantum Mechanics* (Oxford Univ. Press, 1958).

[93] D. Hestess, *Space-Time Algebra* (Gordon and Breach 1966); D. Hestenes and G Sobcyk, *Clifford Algebra to Geometric Calculus* (D. Reidel 1984).

[94] P. Lounesto, *Clifford Algebras and Spinors* (Cambridge Univ. Press 2001).

[95] B. Jancewicz, *Multivectors and Clifford Algebra in Electrodynamics* (World Scientific 1988).

[96] R. Porteous, *Clifford Algebras and the Classical Groups* (Cambridge Univ. Press 1995).

[97] W. Baylis, *Electrodynamics, A Modern Geometric Approach* (Birkhauser 1999).

[98] A. Lasenby and C. Doran, *Geometric Algebra for Physicists* (Cambridge Univ. Press 2002).

[99] C. Castro, *Chaos, Solitons and Fractals* **10**, 295 (1999); **11**, 1663 (2000); **12**, 1585 (2001); C. Castro, *Found. Phys.* **30**, 1301 (2000).

[100] M. Pavšič, *Found. Phys.* **31**, 1185 (2001) [hep-th/0011216].

[101] C. Castro, *Found. Phys.* **30**, 1301 (2000).

[102] S. Ansoldi, A. Aurilia, C. Castro and E. Spallucci, *Phys. Rev. D* **64**, 026003 (2001) [hep-th/0105027].

[103] A. Aurilia, S. Ansoldi and E. Spallucci, *Class. Quant. Grav.* **19**, 3207 (2002).

[104] M. Pavšič, *Found. Phys.* **33**, 1277 (2003) [gr-qc/0211085].

[105] M. Pavšič, *Found. Phys.* **37**, 1197 (2007) [hep-th/0605126].

[106] C. Castro and M. Pavšič, *Prog. Phys.* **1**, 31 (2005).

[107] A. Crumeyrole, *Orthogonal and Symplectic Clifford Algebras* (Dordrecht, 1990)

[108] M. Pavšič, *Phys. Conf. Ser.* **437**, 012006 (2013) [arXiv:1210.6820 [hep-th]].

[109] M. Pavšič, *Adv. Appl. Clifford Algebras*, **23**, 469 (2013) [arXiv:1201.5755 [hep-th]].

[110] M. Pavšič, *J. Phys. Conf. Ser.* **330**, 012011 (2011) [arXiv:1104.2462 [math-ph]].

[111] J. Barbour, *The End of Time* (Oxford Univ. Press, 1999).

[112] V. Fock, *Phys. Z. Sowj.* **12** 404 (1937).

[113] E. C. G. Stueckelberg, *Helv. Phys. Acta* **14**, 322 (1941).

[114] E.C.G. Stueckelberg, *Helv. Phys. Acta* **15**, 23 (1942).

[115] L.P. Horwitz and C. Piron, *Helv. Phys. Acta* **46**, 316 (1973).

[116] L.P. Horwitz and F. Rohrlich, *Phys. Rev. D* **24**, 1528 (1981).

[117] L.P. Horwitz, R.I. Arshansky and A.C. Elitzur, *Found. Phys* **18**, 1159 (1988).

[118] R.P. Feynman, *Phys. Rev* **84**, 108 (1951).

[119] J.R. Fanchi, *Found. Phys.* **23**, 287 (1993), and many references therein.

[120] M. Pavšič, *Found. Phys.* **21**, 1005 (1991).

[121] J.R. Fanchi, *Parametrized Relativistic Quantum Theory* (Kluwer, Dordrecht, 1993).

[122] M. Pavšič *J. Phys. Conf. Ser.* **330**, 012011 (2011) [arXiv:1104.2462 [math-ph]].

[123] L. P. Horwitz, *Relativistic Quantum Mechanics* (Springer, Dordrecht, 2015).

[124] I. Bars, C. Deliduman, and O. Andreev, *Phys. Rev. D* **58**, 066004 (1998); I. Bars, *Phys. Rev. D* **58**, 066006 (1998); I. Bars, *Class. Quant. Grav.* **18**, 3113 (2001); I. Bars, *Phys. Rev. D* **74**, 085019 (2006); I. Bars, and J. Terning, *Extra Dimensions in Space and Time* (Springer, 2010).

[125] P. Zenczykowski, *J. Phys. A* **42**, 045204 (2009) [arXiv:0806.1823 [hep-th]]; P. Zenczykowski, *Int. J. Theor. Phys.* **49**, 2246 (2010). [arXiv:0905.1207 [hep-th]].

[126] C. Castro, *Int. J. Geom. Meth. Mod. Phys.* **6**, 385 (2009); C, Castro, *Int. J. Geom. Meth. Mod. Phys.* **4**, 1239 (2007); C. Castro, *J. Math. Phys.*, **47**, 112301 (2006).

[127] M. Pavšič, *Found. Phys.* **26**, 159 (1996) [gr-qc/9506057].

[128] M. Pavšič, *Phys. Rev. D* **87**, 107502 (2013) [arXiv:1304.1325 [gr-qc]].

[129] T. Padmanabhan, *Gravity and Quantum Theory: Domains of Conflict and Contact* [arXiv:1909.02015 [gr-qc]].

[130] M. Pavšič, *J. Phys. Conf. Ser.* **845**, no.1, 012018 (2017), DOI: 10.1088/1742-6596/8 45/1/012018, Conference: C16-06-06.9 Proceedings.

[131] Y. Ne'eman and E. Elzenberg *Membranes and Other Extendons ("p-Branes")*, World Scientific Lecture Notes in Physics: Volume 39 (1995).

[132] G. Papadopoulos, *Fortschr. Phys.* **44**, 573 (1996).

[133] P. West, *Introduction to Strings and Branes* (Cambridge Univ. Press, 2012).

[134] M. J. Duff, *Benchmarks on the brane*, (2004), [arXiv: hep-th/0407175].

[135] V. A. Rubakov and M.E. Shaposhnikov, *Phys. Lett. B* **125**, 136 1983.

[136] K. Akama, *Lect. Notes Phys.* **176**, 267 (1982) [hep-th/0001113].

[137] M. Visser, *Phys. Lett.* **B159**, 22 (1985).

[138] G. W. Gibbons and D. L. Wiltshire, *Nucl. Phys.* **B287**, 717 (1987).

[139] M. Pavšič, *Phys. Lett. A* **116**, 1 (1986) [gr-qc/0101075].

[140] M. Pavšič, *Nuov. Cim. A* **95**, 297 (1986).

[141] M. Pavšič, *Class. Quant. Grav.* **2**, 869 (1985).

[142] M. Pavšič, *Phys. Lett. A* **107**, 66 (1985).

[143] M. D. Maia, *Phys. Rev. D* **31**, 262 (1985).

[144] M. D. Maia, *Class. Quant. Grav.* **6**, 173 (1989).

[145] V. Tapia, *Class. Quant. Grav.* **6**, L49 (1989).

[146] T. Hori, *Phys. Lett. B* **222**, 188 (1989).

[147] A. Davidson and D. Karasik, *Mod. Phys. Lett. A* **13**, 2187 (1998).

[148] A. Davidson, *Class. Quant. Grav.* **16**, 653 (1999).

[149] A. Davidson, D. Karasik and Y. Lederer, *Class. Quant. Grav.* **16**, 1349 (1999).

[150] M. Pavšič and V. Tapia, *Resource letter on geometrical results for embeddings and branes*, gr-qc/0010045 (2001).

[151] M. Gogberashvili, *Int. J. Mod. Phys. D* **11**, 1639 (2002) [hep-ph/9908347].

[152] L. Randall and R. Sundrum, *Phys. Rev. Lett.* **83**, 4690 (1999) [hep-th/9906064].

[153] M. Pavšč, *Phys. Lett. A* **283**, 8 (2001) [hep-th/0006184].

[154] M. Pavšič, *Grav. Cosmol.* **2**, 1 (1996) [arXiv:gr-qc/9511020].

[155] M. Pavšič, *Found. Phys.* **24**, 1495 (1994).

[156] See, e.g., M. B.Green, J. H. Schwarz and E. Witten *Superstring Theory* (Cambridge Univ. Press, 1987).

[157] M. Kaku, *Introduction to Superstring* (Springer, 1988).

[158] U. Danielson, *Rep. Progr. Phys.* **64**, 51 (2001).

[159] M. Pavšič, "General principles of brane kinematics and dynamics", *Bled Workshops Phys.* **4** 150 (2003).

[160] M. Pavšič, *Int. J. Mod. Phys. A* **31**, 1650115 (2016) [1603.01405 [hep-th]].

[161] A. Schild, *Phys. Rev. D* **16**, 1722 (1977).

[162] B. Rosenstein and L. P. Horwitz, *J. Phys. A: Math, Gen.* **18**, 2115 (1985).

[163] B. U. Rosenstein and M. Usher, *Phys. Rev. D* **36**, 2381 (1987).

[164] E. R. Wagner, B. T. Shields, M. R.Ware, Q. Su and R. Grobe, *Phys. Rev. A* **83**, 062106 (2011).

[165] M. H. Al-Hashimi and U. J. Wiese, *Annals of Physics* **324**, 2599 (2009).

[166] M. Pavšič, *Adv. Appl. Clifford Algebras* **28**, 89 (2018) doi:10.1007/s00006-018-0904-5 [arXiv:1705.02774 [hep-th]].

[167] M. Pavšič, *Mod. Phys. Lett. A* **33**, no. 20, 1850114 (2018) [arXiv:1804.03404 [hep-th]].

[168] R. Jackiw, *Diverse Topics in Theoretical and mathematical Physics* (World Scientific, Singapore, 1995).

[169] T. Newton and E. Wigner, *Rev. Mod. Phys.* **21**, 400 (1949).

[170] M. Pavšič, *Mod. Phys. Lett. A* **34**, 1950186 (2019) [arXiv:1901.01762 [hep-th]].

[171] A. A. Deriglazov, *Phys. Lett. A* **373**, 3920 (2009) [arXiv:0903.1428 [math-ph]].

[172] L. L. Foldy, *Phys. Rev.* **102**, 568 (1956).

[173] M. Pavšič, *Int. J. Mod. Phys. A* **31**, 1650115 (2016) [arXiv:1603.01405 [hep-th]].

[174] T. Blomberg, *Principles of Deductive Theoretical Physics: A Proposal for a General Theory Based on Successive Confidence Estimates on Quantum-Mechanical Wave Functions* (Lambert Academic Publishing, 2018).

[175] D. J. Cirilo-Lombardo, *J. Math. Phys.* **57**, 063503 (2016); doi: 10.1063/1.4953368, [arXiv:1610.03624 [hep-th]].

[176] T. Padmanabhan, *Quantum Field Theory* (Springer, 2016).

[177] T. Padmanabhan, *Eur. Phys. J. C* **78**, 563 (2018) [arXiv:1712.06605 [hep-th]].

[178] S. P. Horwath, D. Schritt and D. V. Ahluwalia, *Amplitudes for space-like separations and causality*, arXiv:1110.1162 [hep-ph].

[179] *Planck units*, Wikipedia, https://en.wikipedia.org/wiki/Planck_units

[180] S. N. Mosley and J. E. G. Farina, *J. Phys. A: Math. Gen.* **23**, 3991 (1990).

[181] G. C. Hegerfeldt, *Phys. Rev. D* **10**, 3320 (1974).

[182] G. C. Hegerfeldt and S. N. M. Ruijsenaars, *Phys. Rev. D* **22**, 377 (1980).

[183] E. Karpov, G. Ordonez, T. Petrosky, I. Prigogine and G. Pronko, *Phys. Rev. A* **62**, 012103 (2000).

[184] N. Barat and J. C. Kimball, *Phys. Lett. A* **308**, 110 (2003).

[185] G. Valente, *Studies in History and Philosophy in Modern Physics* **48**, 147-155 (2014).

[186] R. Haag, *Local Quantum Physics* (Springer-Verlag, Berlin 1996).

[187] H. Reeh and S. Schlieder, *Nuov. Cim.* **22**, 1051 (1961).

[188] I. Antoniou, E. Karpov and G. Pronko, *Found. Phys.* **31**, 1641 (2001).

[189] M. Eckstein, *Phys. Rev. A* **95**, 032106 (2017) [arXiv:1610.00764 [quant-ph]].

[190] G. N. Fleming, *Phys. Rev.* **137**, B188 (1965).

[191] G. N. Fleming, "Lorentz Invariant State Reduction and Localization", in *Proceedings of the Biennial Meeting of the Philosophy of Science Association*, Vol. 1988, Volume Two: Symposia and Invited Papers (1988), pp. 112–126.

[192] S. N. M. Ruijsenaars, *Annals of Phys.* **137**, 33 (1981).

[193] O. M. P. Bilaniuk, V. K. Deshpande and E. C. G. Sudarshan, *Am. J. Phys.* **30** (10), 718 (1962).

[194] G. Feinberg, *Phys. Rev.* **159** (5), 1089–1105 (1967).

[195] G. A. Benford, D. L. Book and W. A. Newcomb, *Phys. Rev. D* **2**, 263 (1970).

[196] Y. Aharonov, A. Komar, L. Susskind, *Phys. Rev.* **182** (5), 1400–1403 (1969).

[197] M. Pavšič, *The extended special theory of relativity*, Preprint, 1971 (Unpublished), http://www-f1.ijs.si/~pavsic/ExtenRel71.pdf

[198] E. Recami and R. Mignani, *Riv. Nuov. Cim.* **4**, 209 (1974).

[199] E. Recami, *Riv. Nuov. Cim.* **9**(6), 1 (1986).

[200] D. Shay and K.L. Miller, *Nuov. Cim. A* **38**, 490 (1977).

[201] A. O. Barut, "Space-Like States in Relativistic Quantum Theory", in *Tachyons, Monopoles, and Related Topics* (Ed. E. Recami, North-Holland, 1978).

[202] V. Vyšín, *Nuov. Cim. A* **40**, 113 (1977).

[203] V. Vyšín, *Nuov. Cim. A* **40**, 125 (1977).

[204] A. F. Antippa, *Nuov. Cim. A* **10**, 389(1972).

[205] B. S. Rajput and O. P. S. Cox, *Phys. Lett. B* **113**, 183 (1982).

[206] V. Majernik, *Found. Phys.* **10**, 357 (1997).

[207] M. C. Pant, P. S. Bisht, O. P. S. Negi, B. S. Rajput, *Can. J. Phys.* **78**, 303 (2000).

[208] R. Mignani and E. Recami, *Lett. Nuov. Cim.* **24**, 171 (1976).

[209] E. A. B. Cole, *Nuov. Cim. A* **40**, 171 (1977).

[210] E A. B. Cole, *J. Phys. A* **13**, 109 (1980).

[211] M. Pavšič, *J. Phys. A* **14**, 3217 (1981).

[212] R. Courant and D. Hilbert, *Methods of Mathematical Physics*, Vol. 1 and 2 (Interscience, London, 1953).

[213] M. Tegmark, *Class. Quant. Grav.* **14**, L69 (1997).

[214] R. P. Feynman, *The Character of Physical Law* (MIT Press, Cambridge, MA, 1965).

[215] H. Everett, *Rev. Mod. Phys.* **29**, 454 (1957).

[216] H. Everett, *On the Foundations of Quantum Mechanics*, Thesis, Princeton University (1956), pp 1–140.

[217] H. Everett, "Theory of the Universal Wavefunction", in *The Many-Worlds Interpretation of Quantum Mechanics* (Eds. B. S. DeWitt and N. Graham, Princeton Univ. Press, 1973), pp. 3–140.

[218] B. S. DeWitt, "Quantum Mechanics and Reality", *Phys. Today* **23**,9, 30 (1970), doi: 10.1063/1.3022331

[219] B. S. DeWitt, "The Many-Universes Interpretation of Quantum Mechanics", in *The Many-Worlds Interpretation of Quantum Mechanics* (Eds. B. S. DeWitt and N. Graham, Princeton Univ. Press, 1973), pp. 167–218.

[220] H. Zeh, *Found. Phys.* **1**, 69 (1970).

[221] D. Deutsch, *The Fabric of Reality* (Penguin Press, London, 1997).

[222] M. Pavšič, *Lett. Nuovo Cim.* **30**, 111 (1981), doi:10.1007/BF02817321

[223] M. Pavšič, *Class. Quant. Grav.* **2**, 869 (1985), [arXiv:1403.6316 [gr-qc]].

[224] D. Deutsch, *Phys. Rev. D* **44**, 3197 (1991).

[225] A.G. Cohen and S.L. Glashow, *Phys. Rev. Lett.* **107**, 181803 (2011).

[226] ICARUS Collaboration, *Physics Letters B* **711**, 270 (2012).

[227] M. Pavšič, *Class. Quant. Grav.* **20**, 2697 (2003) [gr-qc/0111092].

[228] J. Distler and S. Garibaldi, *Commun. Math. Phys.* **298** (2), 419 (2010) [arXiv:0905.2658].

[229] A. G. Lisi, *An Exceptionally Simple Theory of Everything*, arXiv:0711.0770 [hep-th].

[230] T. Adam, et al. (OPERA collaboration) (17 November 2011), *Measurement of the neutrino velocity with the OPERA detector in the CNGS beam*, arXiv:1109.4897v1 [hep-ex].

[231] T. Adam, et al. (OPERA collaboration), *J. High Energy Phys.* **10** (10), 93 (2012) [arXiv:1109.4897 [hep-ex]].

[232] T. Adam, et al. (OPERA collaboration), *J. High Energy Phys.* **1**, 153 (2013) [arXiv:1212.1276 [hep-ex]].

Index

Printed in the United States
by Bookmasters

Printed in the United States
By Bookmasters